W9-BCI-795

The Formula

The Formula

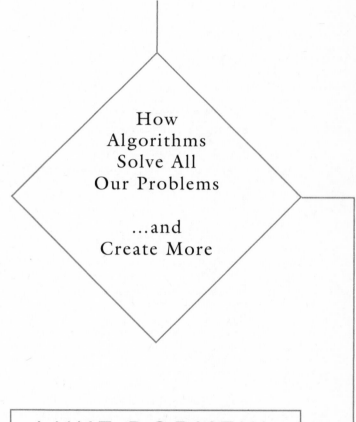

How
Algorithms
Solve All
Our Problems

...and
Create More

LUKE DORMEHL

A PERIGEE BOOK

A PERIGEE BOOK
Published by the Penguin Group
Penguin Group (USA) LLC
375 Hudson Street, New York, New York 10014

USA • Canada • UK • Ireland • Australia • New Zealand • India • South Africa • China

penguin.com

A Penguin Random House Company

THE FORMULA

ISBN: 978-0-399-17053-9

First American edition: November 2014
Previously published in the UK in 2014 by Virgin Books, an imprint of Ebury Publishing.

PRINTED IN THE UNITED STATES OF AMERICA

10   9   8   7   6   5   4   3   2   1

To my friend, Tim Plester

Contents

Acknowledgments

Writing a book is almost always a solitary activity, but I was fortunate enough to be surrounded by a group of people whose love and/or support made *The Formula* a pleasure to work on. Thanks first of all to Clara, Tim and Celia Lunt, as well as members of my family. I could not have completed this project without the invaluable aid of Ed Faulkner, while it would never have got through the door in the first place were it not for my agent Maggie Hanbury, Henry de Rougemont, Simon Garfield and Jake Lingwood. Many thanks to Marian Lizzi, my U.S. editor. Appreciative nods also go in the direction of all those who spent countless hours speaking with me as part of my research (a full list of their names is printed on page 243), in addition to my *FastCo.Labs* editor Chris Dannen, *Cult of Mac*'s Leander Kahney, the excellent Nicole Martinelli, Karl French, tech comms guru Alice Bonasio-Atkinson, Tim Matts, Alex Millington, Michael Grothaus, Tom Atkinson, Simon Callow, and my brothers-from-other-mothers, Andre and Nathan Trantraal. All helped this book along in one way or another. All I can take full credit for are the (hopefully few) mistakes.

An Explanation of the Title, and Other Cyberbole

At their root, algorithms are little more than a series of step-by-step instructions, usually carried out by a computer. However, if their description is straightforward, their inner workings and impact on our lives are anything but.

Algorithms sort, filter and select the information that is presented to us on a daily basis. They are responsible for the search results shown to us by Google, the information about our friends that is highlighted by Facebook, and the type of products Amazon predicts we will be most likely to buy. Increasingly, they will also be responsible for what movies, music and other entertainment look like, which people we are partnered with in predictive relationships, and even the ways in which laws are enforced and police operate. An algorithm can scan through your metadata and recommend that you will likely make a hardworking employee, just as one could accuse you of a crime, or determine that you are unfit to drive a car. In the process, algorithms are profoundly changing the way that we view (to quote Douglas Adams) life, the universe and everything.

One of my favorite observations about technology is the one often attributed to the cultural theorist Paul Virilio: "The invention of the ship was also the invention of the shipwreck." One could, of course, turn this around and say that the inventor of the shipwreck was also the person that invented the ship. Algorithms have had their fair share of shipwrecks (which I will discuss during the course of this book), but they also perform incredibly useful functions: allowing us to navigate through the 2.5 quintillion bytes of data that are generated each day (a million times more information than the human brain is capable of holding) and draw actionable conclusions from it.

As with the old adage about how to carve an elephant statue (you chip away everything that isn't an elephant), I will start out by explaining what this book is not. It is not, for one thing, a computer science textbook about algorithms. There are far better books (and, indeed, far more qualified writers) to achieve this task.

Neither is it a history of the algorithm as a concept. While I considered attempting such a thing, I was put off by both the sheer scale of the project and the fact that its end result—while no doubt fascinating under the stewardship of the right author— would be not entirely dissimilar to the textbook I also shied away from. By this I do not mean that a history book and a textbook are necessarily the same thing, but rather that a history of a once-niche mathematical concept would likely appeal only to those mathematicians or computer scientists already familiar with it.

Instead, I want to tell the story of the myriad ways (some subtle, others less so) that algorithms affect all of our lives: from the entertainment we enjoy to the way we think about human relationships. What do scoring hot dates, shooting Hollywood turkeys, bagging your own poo, and cutting opportunities for lawyers' fees have in common? This is a book about the algorithmization of life as we know it.

In my day job, writing about a field known as the "digital humanities" for *Fast Company*, I'm constantly considering the implications of "algorithmic" culture and the idea (not always a misguided one) that no matter what the problem, it can be solved with the right algorithm.

A typical illustration of what I mean can be seen in Bill Tancer's 2009 book *Click: What We Do Online and Why It Matters*. Tancer—described in at least one place as "the world's preeminent expert on online [behavior]"—begins his book by describing a radio interview he listened to in the car one day. Being interviewed was a British psychologist referring to a mathematical formula he had developed to determine the most objectively depressing week of the year. After much work, he had discovered that this was the third week of January, a feat brought about by the convergence of failed New Year's resolutions, credit-card debt accumulated over the holiday season, and the usual dismal weather patterns. Tancer notes that he remained unconvinced: a perspective that was later backed up when the formula was severely criticized for its lack of scientific rigor. However, his lack of conviction has nothing to do with the suggestion that a reductive formula could possibly provide answers on a topic as complex and multifaceted as depression, but rather because he believes that he had come up with a better formula.[1]

In other words, his problem wasn't with the existence of the sum, but rather with its working.

This book was spurred by years of hearing similar observations, all claiming that there is no problem technology cannot reduce to its most formulaic level and thereby determine objective answers in response to. This thinking is the reason "The Formula" is in upper case rather than simply existing as a catchall for the various technical processes I describe. It implies an ideological element, and that ideology is evidenced by the more expansive view I take of algorithms and their associated technological apparatus: conceiving of them as the embodiment of a particular form of techno-rationality, symptomatic of a type of social ordering built around the promise of objectivity. In this way I use The Formula much as the late American political scientist and communications theorist Harold Lasswell used the word "technique": referring, in Lasswell's words, to "the ensemble of practices by which one uses available resources to achieve values." It is about both the application and the scope of application, as well as the existence of objective truths lurking just beneath the surface—to be teased out with the right data-mining tools.

Writers on technology tend, with a few notable exceptions, to be overwhelmingly utopian in their outlook. To them, all progress is positive. As a result, there is a tendency among technology writers to christen each new invention as the totemic figurehead of its own "era"—something that has led to the disdainful term "cyberbole." While this book could very well join the number of volumes about algorithms and big data already lining the shelves, what I am interested in goes back much further than simply the birth of the Internet or the age of the personal computer.

Writing during the first half of the 1960s, the French sociologist (and Christian anarchist!) Jacques Ellul described a creature known as the Technical Man, an individual "fascinated by results, by the immediate consequences of setting standardized devices into motion . . . committed to the never-ending search for 'the one best way' to achieve any designated objective." This objective could occasionally be clouded (or else speeded up) by a naive enthusiasm for the means of getting there: not by anything so unquantifiable as ethical concerns, but rather by an enthusiasm for the ingenuity, elegance and "spectacular effectiveness" of man's ability to dream up solutions.

As Ellul's observation proves, this approach is not therefore a new one, and the founders of Google and the heads of the various high-tech companies I discuss are not the first people to display what the late American sociologist Lewis Mumford called the "will-to-order"—meaning the desire to make formulaic sense of the world. Writing in the 1930s, long before the birth of the modern computer, Mumford noted that automation was simultaneously for "enlarging the mechanical or sensory capacities of the human body" and "for reducing to a measurable order and regularity the processes of life." To make sense of a big picture, we reduce it, he suggested. To take an abstract concept such as human intelligence and turn it into something quantifiable, we abstract it further, stripping away complexity and assigning it a seemingly arbitrary number, which becomes a person's IQ.

What is new is the scale that this idea is now being enacted upon, to the point that it is difficult to think of a field of work or leisure that is not subject to algorithmization and The Formula. This book is about how we reached this point, and how the age of the algorithm impacts and shapes subjects as varied as human creativity, human relationships (and, more specific-ally, romantic relationships), notions of identity and matters of law.

Algorithms are very good at providing us with answers in all of these cases.

The real question is whether they give the answers we want.

CHAPTER 1

The Quantified Selves

Larry Smarr weighed close to 200 pounds when he arrived in La Jolla, California, in 2000. The photograph from his driver's license at the time depicts an overweight 51-year-old with a soft, round face and the fleshy ripple of a double chin. Although he regularly tested his mind as a leading physicist and expert in supercomputing, Smarr had not exercised his body properly in years. He regularly drank Coke and enjoyed chowing down on deep-fried, sugar-coated pastries.[1] Moving to the West Coast to run a new institute at the University of California called the California Institute for Telecommunications and Information Technology, he was suddenly confronted with a feeling he hadn't experienced in years: a sense of deep inadequacy. "I looked around and saw all these thin, fit people," he remembers. "They were running, bicycling, looking beautiful. I realized how different I was."

Smarr's next question was one shared by scientists and philosophers alike: Why? He visited his local bookshop and bought "about a zillion" diet books. None of them was deemed satisfactory to a man whose brain could handle the labyrinthine details of supernovas and complex star formations, but for whom healthy eating and regular exercise seemed, ironically enough, like complex astrophysics. "They all seemed so arbitrary," he says. It wasn't until he discovered a book called *The Zone*, written by biochemist Barry Sears, that Smarr found what it was that he was looking for. Sears treated the body as a coupled nonlinear system in which feedback mechanisms like the glucose-insulin and immune systems interact with one another. That was an approach that Smarr could relate to. Inspired, he started measuring his weight, climbing naked onto a pair of scales each day and writing down the number that appeared in front of him. Next, he hired a personal trainer and began keeping track of the amount of exercise he participated in on a daily basis. After that it was on to diet—breaking food down into its "elemental units" of protein, fat, carbohydrates, fiber, sugar and salt, and modifying his diet to remove those inputs that proved detrimental to well-being. "Think of it like being an engineer, reverse-engineering the subsystems of a car," Smarr says. "From that you can derive that you need a certain level of petrol in order to run, and that if you put water in your gas tank you will tear apart the car. We don't think that way about our bodies, but that is the way we ought to think."

It didn't take long until Smarr turned to more complex forms of technology to help him lose weight. He purchased and began wearing Polar WearLink heart-rate straps, FitBits, BodyMedia, and other pieces of wearable tech that use algorithms to convert body metrics into data. Wanting to check his progress, Smarr started paying for blood tests at a private laboratory and then—in a quest for yet more numbers to pore over—began FedEx-ing off his stool for regular analysis. "I didn't have a biology or medical background, so I had to teach myself," Smarr says of his process.

One of the numbers he wound up fixating on related to complex reactive proteins, which act as a direct measure of inflammation in the body. In a normal human body this number should be less than one. In Smarr's case it was five. Over time it rose to 10, then 15. As a scientist, he had discovered a paradox. "How was it that I had reduced all of the things that normally drive inflammation in terms of my food supply, but the numbers were growing and growing with this chronic inflammation?" he muses. "It didn't make any sense."

At that point, Smarr decided to visit his doctor to present the findings. The appointment didn't go as planned.

"Do you have any symptoms?" the doctor asked.

"No," Smarr answered. "I feel fine."

"Well, why are you bothering me, then?"

"Well, I've got these great graphs of my personal data."

"Why on earth are you doing that?" came the response.

The doctor told Smarr that his data was too "academic" and had no use for clinical practice. "Come back when there's something actually wrong with you, rather than just anomalies in your charts," the doctor said.

Several weeks later, Smarr felt a severe pain in the left side of his abdomen. He went back to the doctor's and was diagnosed with diverticulitis, a disease caused by acute inflammation. It was the perfect illustration of Smarr's problem: doctors would deal only in clinical symptoms, unwilling to delve into the data that might actually prove preventative. Having learned an important lesson, Smarr decided to take over his own health tracking.

"People have been brainwashed into thinking that they have no responsibility for the state of their bodies," he says. "I did the calculation of the ratio of two 20-minute doctor visits per year, compared to the total number of minutes in the year, and it turns out to be one in 10,000. If you think that someone is going to be able to tell you what's wrong with you and fix the problem in one 10,000th of the time that you have available to do the same, I'd say that's the definition of insanity. It just doesn't make any sense."

In the aftermath of the doctor's visit, Smarr began obsessively tracking any and all symptoms he noticed and linking each of these to fluctuations in his body data. He also upped the amount of information he was looking at, and started using complex data-mining algorithms to sift through it looking for irregularities. Another high number he zeroed in on referred to lactoferrin, an antibacterial agent shed by white blood cells when they are in attack mode, a bit like a canister of tear gas being dispersed into a crowd of people. This number was meant to be less than seven. It was 200 when Smarr first checked it, and by May 2011 had risen to 900. Searching through scientific literature, Smarr diagnosed himself as having a chronic autoimmune disorder, which he later narrowed down to something called Crohn's disease. "I was entirely led there by the biomarkers," he notes.

In this sense, Smarr is the epitome of algorithmic living. He tracks 150 variables on a constant basis, walks 7,000 steps each day, and has access to millions of separate data points about himself. As such, both his body and his daily life are divided, mathematized and codified in a way that means that he can follow what is happening inside him in terms of pure numbers. "As our technological ability to 'read out' the state of our body's main subsystems improves, keeping track of changes in our key biochemical markers over time will become routine, and deviations from the norm will more easily reveal early signals of disease development," Smarr argues in an essay entitled "Towards Digitally Enabled Genomic Medicine: A 10-Year Detective Story of Quantifying My Body."[2]

Using his access to the University of California's supercomputers, Smarr is currently working on creating a distributed planetary computer composed of a billion processors that—he claims within ten years—will allow scientists to create working algorithmic models of the human body. What, after all, is the body if not a complex system capable of tweaks and modifications?

As journalist Mark Bowden observed in a 2012 article for the *Atlantic*, "If past thinkers leaned heavily on the steam engine as an all-purpose analogy—e.g., contents under pressure will explode (think Marx's ideas on revolution or Freud's about repressed desire)—today we prefer our metaphors to be electronic."

And when it comes to symbolic representations, many people (Smarr included) prefer formulas to metaphors.[3]

Self-Knowledge Through Numbers

Fitting an "$n = 1$" study in which just one person is the subject, Larry Smarr's story is exceptional. Not everyone is an expert in supercomputing and not everyone has the ability, nor the resources (his regimen costs between $5,000 and $10,000 each year) to capture huge amounts of personal data, or to make sense of it in the event that they do.

But Smarr is not alone. As a data junkie, he is a valued member of the so-called Quantified Self movement: an ever-expanding group of similar individuals who enthusiastically take part in a form of self-tracking, somatic surveillance. Founded by *Wired* magazine editors Gary Wolf and Kevin Kelly in the mid-2000s, the Quantified Self movement casts its aspirations in bold philosophical terms, promising devotees "self-knowledge through numbers."[4] Taking the Positivist view of verification and empiricism, and combining this with a liberal dose of technological determinism, the Quantified Self movement begs the existential question of what kind of self can possibly exist that is unable to be number-crunched using the right algorithms?

If Socrates concluded that the unexamined life was not worth living, then a 21st-century update might suggest the same of the unquantified life. As with René Descartes' famous statement, *cogito ergo sum* ("I think, therefore I am")—I measure, therefore I exist.

In a sense, "Selfers" take Descartes' ideas regarding the self to an even more granular level. Descartes imagined that consciousness could not be divided into pieces in the way that the body can, since it was not corporeal in form. Selfers believe that a person can be summarized effectively so long as the correct technology is used and the right data gathered. Inputs might be food consumed or the quality of surrounding air, while states can be measured through mood, arousal and blood oxygen levels, and performance in terms of mental and physical well-being.

"I like the idea that someone, somewhere is collecting all of this data," says Kevin Conboy, the creator of a quantified sex app, Bedpost, which I will return to later on in this book. "I have this sort of philosophical hope that these numbers exist somewhere, and that maybe when I die I'll get to see them. The idea that computer code can give you an insight into your real life is a very powerful one."

A typical Quantified Self devotee (if there is such a thing) is "Michael." Every night, Michael goes to bed wearing a headband sensor. He does this early, because this is when the sensor informs him that his sleep cycles are likely to be at their deepest and most restorative. When Michael wakes up he looks at the data for evidence of how well he slept. Then he gets up, does some push-ups and meditates for a while, before turning on his computer and loading a writing exercise called "750 Words" that asks him to write down the first 750 words that come to mind.[5] When he has done this, text-analysis algorithms scour through the entry and pull up revealing stats about Michael's mood, mind-set and current preoccupations—some of which he may not even be consciously aware he is worrying about. After this, he is finally ready to get moving (using a FitBit to monitor his steps, of course). If he doesn't carry out these steps, he says, "I'm off for the rest of the day."[6]

Robo-cize the World

While QS's reliance on cutting-edge technology, social networking and freedom-through-surveillance might seem quintessentially modern—very much a creation of post-9/11 America—the roots of what can be described as "body-hacking" go back a number of years. The 1980s brought about the rise of the "robo-cized" athletes who used Nautilus, Stairmaster and other pieces of high-tech gym equipment to sculpt and hone their bodies to physical perfection. That same decade saw the advent of the port-able technology known as the Sony Walkman (a nascent vision of Google Glass to come), which transformed public spaces into a controllable private experience.[7] Building on this paradigm, the 1990s was home to MIT's Wearable Computing Group, who took issue with what they considered to be the premature usage of the term "personal computer" and insisted that:

> A person's computer should be worn, much as eyeglasses or clothing are worn, and interact with the user based on the context of the situation. With heads-up displays, unobtrusive input devices, personal wireless local area networks, and a host of other context sensing and communication tools, the wearable computer can act as an intelligent assistant, whether it be through a Remembrance Agent, augmented reality, or intellectual collectives.[8]

There appear to be few limits to what today's Quantified Selfers can measure. The beauty of the movement (if one can refer to it in such aesthetic terms) is the mass customization that it makes possible. By quantifying the self, a person can find apparently rigorous answers to questions as broad or spe-cific as how many minutes of sleep are lost each night per unit of alcohol consumed, how consistent their golf swing is, or whether or not they should stay in their current job. Consider, for example, the story of a young female member of the Quan-tified Self movement, referred to only as "Angela."

Angela was working in what she considered to be her dream job, when she downloaded an app that "pinged" her multiple times each day, asking her to rate her mood each time. As patterns started to emerge in the data, Angela realized that her "mood score" showed that she wasn't very happy at work, after all. When she discovered this, she handed in her notice and quit.

"The one commonality that I see among people in the Quantified Self movement is that they have questions only the data can answer," says 43-year-old Selfer Vincent Dean Boyce. "These questions may be very simplistic at first, but they very quickly become more complex. A person might be interested in knowing how many miles they've run. Technology makes that very easy to do. A more advanced question, however, would be not only how many miles a person has run, but how many other people have run the same amount? That's where the data and algorithms come in. It's about a quest for knowledge, a quest for a deeper understanding not only of ourselves, but also of the world we live in."

Boyce has always been interested in quantification. As a New York University student enrolled in the Interactive Tele-communications Program, he once attached some sensors, micro-controllers and an accelerometer to a model rocket and launched it off the ground. "What was interesting," he says, "is that I was able to make a self-contained component that could be sent somewhere, that could gather information, and that I could then retrieve and learn something about." After analyzing the rocket's data, Boyce had his "Eureka!" moment. A lifelong skateboarder and surfer, he decided to attach similar sensors to his trusty skateboard and surfboard to measure the mechanical movements made by each. He also broke the surrounding environment down into hundreds of quantifiable variables, ranging from weather and time of day to (in the case of surfing) tidal changes and wave height. "From a Quantified Self standpoint," Boyce notes, "I can . . . think about where it was that I surfed from a geospatial type of framework, or what equipment I was using, what the conditions were like . . . [It] represents me doing something in space and time."

In this way, Selfers return to the romantic image of the rugged individualists of the American frontier: an image regularly drawn upon by Silicon Valley industrialists and their followers. The man who tracks his data is no different from the one who carves out his own area of land to live on, who draws his own water, generates his own power, and grows his own food. In a world in which user data and personal information is gathered and shared in unprecedented quantities, self-tracking represents an attempt to take back some measure of control. Like Google Maps, it puts the individual back at the center of his or her universe. "My belief is that this will one day become the norm," Boyce says of the Quantified Self. "It will become a commodity, with its own sense of social currency."

Shopping Is Creating

One of the chapters in Douglas Coupland's debut novel *Generation X*, written at the birth of the networked computer age, is titled "Shopping Is Not Creating."[9] It is a wonderfully pithy observation about 1960s activism sold off as 1990s commercialism, from an author whose fiction books *Microserfs* and *JPod* perfectly lampoon techno-optimism at the turn of the millennium. It is also no longer true. Every time a person shops online (or in a supermarket using a loyalty card) their identity is slightly altered, being created and curated in such a way that is almost imperceptible.

This isn't just limited to shopping, of course. The same thing happens whenever you open a new web-browsing window and surf the Internet. Somewhere on a database far away, your movements have been identified and logged. Your IP address is recorded and "cookies" are installed on your machine, enabling you to be targeted more effectively with personalized advertisements and offers. Search regularly for news on a particular sport and you will begin to spot adverts related to it wherever you look—like the murderer played by Robert Walker in Hitchcock's *Strangers on a Train*, who sees constant reminders of the woman he killed. Mention the words "Cape Town" in an e-mail, for instance, and watch the flood of "Cheap flights to South Africa" messages flood in.

It was the American philosopher and psychologist William James who observed, in volume one of his 1890 text *The Principles of Psychology*, that "a man's self is the sum total of all that he [can] call his, not only his body and his psychic powers, but his clothes and his house, his wife and children, his ancestors and friends, his reputation and works, his lands, and yacht and bank account."[10] This counts for double in the age of algorithms and The Formula. Based on a person's location, the sites that they visit and spend time on, and the keywords that they use to search, statistical inferences are made about gender, race, social class, interests and disposable income on a constant basis. Visiting *Perez Hilton* suggests a very different thing from *Gizmodo*, while buying airline tickets says something different from buying video games. To all intents and purposes, when combined, these become the algorithmic self: identity and identification shifted to an entirely digital (and therefore measurable) plane.

Your Pleasure Is Our Business

Identity is big business in the age of The Formula. The ability to track user movements across different websites and servers has led to the rise of a massive industry of web analytics firms. These companies make it their mission not only to amass large amounts of information about individuals, but also to use proprietary algorithms to make sense of that data.

One of the largest companies working in this area is called Quantcast. Headquartered in downtown San Francisco—but with additional offices in New York, Dublin, London, Detroit, Atlanta, Chicago and Los Angeles—Quantcast ranks among the top five companies in the world in terms of measuring audiences, having raised in excess of $53.2 million in venture capital funding since it was founded in 2006. Its business revolves around finding a formula that best describes specific users and then advising companies on how to best capitalize on this. "You move away from the human hypothesis of advertising," explains cofounder Konrad Feldman, "where someone theorizes what the ideal audience for a product would be and where you might be able to find these people—to actually measuring an advertiser's campaign, looking at what's actually working, and then reverse-engineering the characteristics of an audience by analyzing massive quantities of data."

Before starting Quantcast, English-born University College London graduate Feldman founded another business in which he used algorithms to detect money laundering for some of the world's leading banks. "We looked through the billions of transactions these banks deal with every month to find suspicious activity," he says. It was looking at fraud that made Feldman aware of the power of algorithms' ability to sort through masses of data for patterns that could be acted upon. "It could represent anything that people were interested in," he says excitedly. "Finances were interesting data, but it only related to what people spend money on. The Internet, on the other hand, has information about interests and changes in trends on the macro and micro level, all in a single data format." He was hooked.

"Historically, measurement was done in retrospect, at the aggregate level," Feldman says of the advertising industry. "That's what people understood: the aggregate characteristics of an audience." When Feldman first moved to the United States, he was baffled by the amount of advertising on television, which often represented 20 minutes out of every hour. It was a scatter-gun approach, rather like spraying machine-gun bullets into a river and hoping to hit individual fish. Whatever was caught was more or less done so by luck. Of course, a television channel can't change it for every viewer, Feldman explains. The Internet, however, was different. Much like the customized user recommendations on Amazon, Quantcast's algorithmically generated insights meant that online shopkeepers could redecorate the shop front for each new customer. In this way, audiences are able to be divided into demographics, psychographics, interests, lifestyles and other granular categories. "Yep, we're almost psychic when it comes to reading behavior patterns and interpreting data," brag Quantcast's promotional materials. "We know before they do. We know before you do. We can tell you not only where your customers are going, but how they're going to get there, so we can actually influence their paths."

Quantcast's way of thinking is rapidly becoming the norm, both online and off. A Nashville-based start-up called Facedeals promises shops the opportunity to equip themselves with facial recognition-enabled cameras. Once installed, these cameras allow retailers to scan customers and link them to their Facebook profiles, then target them with personalized offers and services based upon the "likes" they have expressed online. In late 2013, UK supermarket giant Tesco announced similar plans to install video screens at its checkouts around the country, using inbuilt cameras equipped with custom algorithms to work out the age and gender of individual shoppers. Like loyalty cards on steroids, these would then allow customers to be shown tailored advertisements, which can be altered over time, depending on both the date and time of day, along with any extra insights gained from monitoring purchases. "It is time for a step-change in advertising," said Simon Sugar, chief executive of Amscreen, who developed the OptimEyes technology behind the screens. "Brands deserve to know not just an estimation of how many eyeballs are viewing their adverts, but who they are, too."[11]

The Wave Theory

This notion of appealing to users based on their individual surfing habits taps—ironically enough—into the so-called wave theory of futurist Alvin Toffler.[12] In his 1980 book *The Third Wave*, Toffler described the way in which technology develops in waves, with each successive wave sweeping aside older societies and cultures.[13] There have been three such waves to date, Toffler claimed. The first was agricultural in nature, replacing the hunter-gatherer cultures and centering on human labor. The second arrived with the Industrial Revolution, was built around large-scale machinery, and brought with it the various "masses" that proliferated in the years since: mass production, mass distribution, mass consumption, mass education, mass media, mass recreation, mass entertainment and weapons of mass destruction. The Third Wave, then, was the Information Age, ushering in a glorious era of "demassification" under which individual freedoms could finally be exercised outside the heaving constraints of mass society. Demassification would, Toffler argued, be "the deepest social upheaval and creative restructuring of all time," responsible for the "building [of] a remarkable new civilization from the ground up." And it was all built on personalization.

Please Hold to Be Connected to Our Algorithm

It is well known that not every call-center agent is equipped to handle every type of call that comes in. The larger the company, the less likely it is that any one person will be able to deal with every single inquiry, which is the reason customers are typically routed to different departments in which agents are trained to have different skills and knowledge bases. A straightforward example might be the global company whose call centers regularly receive calls in several different languages. Both callers and agents may speak one or more of several possible languages, but not necessarily all of them. When the French-speaking customer phones up, they may be advised to press "1" on their keypad, while the English-speaking customer might be instructed to press "2." They are then routed through to the person best suited to deal with their call.

But what if—instead of simply redirecting customers to different call-center agents based upon language or specialist knowledge—an algorithm could be used to determine particular qualities of the person calling in: based upon speech patterns, the particular words they used, and even details as seemingly trivial as whether they said "um" or "err"—and then utilize these insights to put them through to the agent best suited for dealing with their *emotional* needs?

Chicago's Mattersight Corporation does exactly that. Based on custom algorithms, Mattersight calls its business "predictive behavioral routing." By dividing both callers and agents into different personality types, it can make business both faster and more satisfactory to all involved. "Each individual customer has different expectations and behaviors," Mattersight notes in promotional materials. "Similarly, each individual employee has different strengths and weaknesses handling different types of calls. As a result, the success of a given customer interaction is often determined by which employee handles that interaction and how well their competencies and behavioral characteristics align with each specific customer's needs."

The man behind Mattersight's behavioral models is a clinical psychologist named Dr. Taibi Kahler. Kahler is the creator of a type of psychological behavioral profiling called Process Communication. Back in the early 1970s, Kahler interned at a private psychiatric hospital. While he was there, he created something called a "Miniscript" based on his observations about patients in distress. The work wound up winning him the 1977 Eric Berne Memorial Scientific Award. What Kahler noticed was that certain predictable signs precede particular incidents of distress, and that these distress signs are linked to specific speech patterns. These, in turn, led to him developing profiles on the six different personality types he saw recurring. The personality types are as follows:

Personality type	Personality traits	How common?
"Thinkers"	Thinkers view the world through data. Their primary way of dealing with situations is based upon logical analysis of a situation. They have the potential to become humorless and controlling.	1 in 4 people
"Rebels"	Rebels interact with the world based on reactions. They either love things or hate them. Many innovators come from this group. Under pressure they can be negative and blameful.	1 in 5 people
"Persisters"	Persisters filter everything through their opinions. Everything is measured up against their worldview. This describes the majority of politicians.	1 in 10 people
"Harmonizers"	Harmonizers deal with everything in terms of emotions and relationships. Tight situations make this group overreactive.	3 in 10 people
"Promoters"	Promoters view everything through action. These are the salesmen of the world, always looking to close a deal. They can be irrational and impulsive.	1 in 20 people
"Imaginers"	Imaginers deal in unfocused thought and reflection. These people operate in vivid internal worlds and are likely to spot patterns where others cannot.	1 in 10 people

Although everyone has all six personality types to a greater or lesser degree, people will respond best to individuals who reflect their own primary personality type. If people's communication needs are not met by being given the kind of positive "feedback" they require (a feelings-oriented person being asked cold hard facts, for example) they go into distress, which can be diffused only if the person on the other end of the conversation is able to adequately pick up on the warning signals and respond appropriately.

In a call-center environment this knowledge results in an extraordinary qualitative change, according to Mattersight. A person patched through to an individual with a similar personality type to their own will have an average conversation length of five minutes, with a 92 percent problem-resolution rate. A caller paired up to a conflicting personality type, on the other hand, will see their call length double to ten minutes—while the problem-resolution rate tumbles to 47 percent.

Process Communication isn't only being used by Mattersight, however. In the past, Kahler has helped NASA develop algorithms to aid with the selection of its astronauts, since his model can accurately predict the personality types that won't crack under the high-pressure atmosphere of space travel. ("Persisters"—who strive for perfection and encourage others to reach their peak performance—prove to be the best personality fit.) Kahler's company, Kahler Communications, also has a number of ongoing projects designed to help organizations come up with data-driven and algorithmic solutions to questions related to personality.

"From our perspective this is the key to diversity," says Robert Wert, a former attorney who was employed as the COO of Kahler Communications when I had the opportunity to speak with him. "If all cultures are made up of the same building blocks, all of whom have the same type of interactions both positive and negative, then the real diversity is in personality type. It's not in ethnicity, it's not in gender, it's not in anything else. I see this as the great equalizer. If you can walk into a room and immediately start speaking to someone who's of a different background to you, and you can identify the same traits in them that you've dealt with for the rest of your life, that person is no longer the Other."

The Lake Wobegon Strategy

Founded in 2011, Gild is a recruitment company that serves some of the tech industry's biggest and best-known players. Currently focused on automating the discovery of talented programmers, Gild's mission statement is to apply The Formula to the notoriously unreliable hiring process. To do this, the company uses algorithms to analyze individuals on tens of thousands (soon hundreds of thousands) of different metrics and data points—mining them for insights in what Gild refers to as "broad predictive modeling."

The success stories the company trots out are impressive. A typical one tells of 26-year-old college dropout Jade Dominguez, who lived off an increasing line of credit-card debt in South Pasadena, California, while teaching himself computer programming.[14] After being "discovered" by Gild's algorithm, he now works as a programmer at the company that found him. His story is hardly unique, either. "These are people whose CVs you wouldn't look twice at, but who our algorithm predicts would be perfect for the job," says Vivienne Ming,

Gild's chief scientist. "For some of our customers, that is exactly what they're looking for. These are companies that are flooded with résumés. They don't need us to find people; they need us to find *different* people."

The first time I spoke with Ming, it was May 2013, and she was sitting in the back of a taxicab on her way to San Francisco International Airport. A tall, striking woman with silver-blue eyes and strawberry-blond hair, Ming is a theoretical neuroscientist with a Carnegie Mellon University pedigree. Effortlessly assured, her geeky engineering side is evidenced by the fact that she wears a prerelease Google Glass headset. In addition to her neuroscience background, Ming's Twitter profile describes her as an "intrepid entrepreneur, undesirable superhero [and] very sleepy mother."

Ming is deeply invested in Gild's utopian vision of turning the workplace into the kind of meritocracy she believes it should be. "This is the way things ought to work, right?" she says, rhetorically. "The person making the hiring decisions really should have an accurate picture of who I am—not just a snap judgment made because I look a certain way. But believe me, the way that people look is a huge influence on hiring."

If there is a reason why the idea of people being misjudged on first appraisal hits home particularly hard with Ming it may have something to do with her background. Born Evan Campbell Smith, Ming underwent gender reassignment surgery in 2008, having "ghosted [her] way through life" up until that point. "I was not a classically good student," she explains. "I frequently failed my classes, I was constantly in trouble at school. I was not engaged, but I deeply cared about the learning experience. The most trouble I ever got in was lying to stay in an honors chemistry course. I loved being there. I loved learning." After Ming underwent the operation to become a woman, she noticed that she was treated differently: being asked fewer questions about math than she had as a man, and not being invited to so many social events by male colleagues and business connections.

To Ming, there exist two main problems with classic hiring strategies. The first is that they are inherently biased. While the majority of people appreciate the value of recruiting people with a different background from themselves, they are often not exposed to these individuals in social settings. Why, she asks, is a typical start-up composed of similar-looking individuals of approximately the same age, with the same scruffy engineering look? Because they hired people they knew. The person who is good friends with a white, male, upper-middle-class, hardworking engineer is statistically more likely to be a hardworking engineer themselves. They are also likely to be white, male and upper middle class. As data-driven algorithmic culture has taken over, these casual assumptions have in many cases become codified. To get a job at Facebook, one of the initial tests used in the weeding-out process is to find a person already working for Facebook who knows you. This is the same idea as LinkedIn, whose algorithms search for connections between an individual and the person they are trying to meet. Although the idea is certainly neat on one level, it can also have the unfortunate effect of excluding a significant number of people from diverse regional, social and cultural backgrounds.

The other problem that Ming explains (and to a scientist this is almost certainly worse) is that previous hiring strategies have proven inaccurate when it comes to forecasting who will succeed

in a workplace role. In a place like Silicon Valley, where the supposed objectivity of data-driven hiring is prized above all else,
this is particularly unforgivable. Google, for example, employs
what it calls the Lake Wobegon Strategy for hiring—named
after American humorist Garrison Keillor's claim that he grew
up on the fictitious Lake Wobegon, where "all the women are
strong, all the men are good-looking, and all the children are
above average." According to Google's Lake Wobegon Strategy,
to maintain a high level of skill in an organization that is dou-
bling in size each year, new employees should be above the mean
skill level of current Googlers in order to be hired. To measure
something as unquantifiable as "skill," Google traditionally
placed a strong emphasis on academic results. A person's GPA
and university were considered a strong predictor of workplace
success, since they showed past evidence of rigor, stickability and
the ability to meet deadlines. An individual who studied computer science at MIT might not be the best computer scientist in
the world, but it is surely safe to assume that they are at least
"good enough" to have gotten into the course in the first place.
Pick a random person who *didn't* go to MIT, on the other hand,
and while there is still the chance that they will be brilliant, the
likelihood that they will be terrible is far higher. In a risk-averse
industry where people are rarely given a bonus for betting on
the long shot that pays off—but could very easily lose their job
for hiring a person deemed unsuitable for a particular role—it
is no wonder that many high-tech companies would choose to
play it safe.

    As datasets piled up, however, and algorithms began scour-
ing that information for patterns, Google realized that the metrics they were using to predict job performance (including

school grades, exam results, previous job experience and even face-to-face interviews) offered very little in the way of accuracy when forecasting who it was that was likely to excel in a particular position. This in itself was not an entirely new revelation. In the 1960s, telecommunications giant AT&T conducted IQ tests on low-level managers and then followed them for the next 20 years of their career to see how each employee progressed within the organization. In line with popular wisdom, AT&T's assumption was that those individuals with a higher IQ would rise to the highest level, while those with lower IQs would settle lower down in the company, like water that finds its own level. Instead, what was discovered was that IQ scores explained less than 15 percent of the variance between managers in terms of career achievement. The rest was an unmeasurable combination of personality traits, emotional attributes, sociability and a number of other characteristics that can determine success.

Once Google realized the need to open up the parameters of what it looked for in a new employee, the cultural space was cleared for Gild. Instead of hiring from a population of tens of thousands each year, high-tech companies now have a population consisting of almost everybody to choose from. After all, it is never knowable where the diamond in the rough might pop up. To hedge bets, a prospective employee being looked over by Gild is measured on practically every scrap of information in existence about them. Vivienne Ming likens the difference between this approach and that of a human recruiter to a human chess player competing against Deep Blue—the supercomputer that famously defeated grand master Garry Kasparov in what *Newsweek* described as "The Brain's Last Stand." "The computer is plotting every possible move and choosing the optimal move," she says. "Chess experts are not. They have implicitly discarded the vast majority of possible moves and are only considering two, three, four possibilities. They just happen to be great ones." That's also true of people making hiring decisions, Ming suggests—only that the metrics considered in this case don't happen to be so great. By looking at as many data points as possible about a person, anomalous factors like whether a person being interviewed was having an off day are bypassed. Gild additionally looks at where individuals spend time online, since this has been shown to be a strong predictor of workplace skills. "If you spend a lot of time blogging it suggests that you're not quite as good a programmer as someone who spends their time on Quora," Ming says, referring to the question-and-answer website founded by two former Facebook employees. Even Twitter feeds are mined for their insights, using semantic and sentiment analysis. At the end, factors are combined to give prospective employees a "Gild Score" out of 100.

"It's very cool if you're geeky about algorithms, but the really important take-away is that what we end up with is truly independent dimensions for describing people out in the world," she says. "We're talking about algorithms whose entire intent and purpose is to aggregate across your entire life to build up a very accurate representation of who you are."

Quantifying Human Potential

Gild is not the only interested party looking to open up the number of metrics individuals are judged on in the workplace. In 2012, three universities carried out a study as to whether or not Facebook profiles can be used to predict how successful a person is likely to be at their job. By analyzing photos, wall posts, comments and profiles, researchers argued that questions such as "Is this person dependable?" and "How emotionally stable is this person?" can be answered with a high level of certainty. Favorable evaluations were given to those students who had traveled, had more friends and demonstrated a wide range of hobbies and interests. Partying photos didn't necessarily count as negative either, since people depicted as partiers were characterized as extroverted and friendly, both viewed as ideal qualities for workplace success. Six months after making their initial predictions, the study's authors followed up with the employers of their 56 test subjects and found a strong correlation between job performance and the Facebook scores that had been awarded for traits such as conscientiousness, agreeability and intellectual curiosity. Their conclusion was that Facebook profiles are strong predictors since candidates will have a harder time "faking . . . their personalities" on a social network than they would in a conventional job interview.[15]

This attitude toward quantification and statistical analysis is one that is heard regularly from exponents of The Formula. It owes its biggest debt to the Belfast-born mathematical physicist and engineer Lord Kelvin, who suggested that the thing that cannot be measured cannot possibly be improved. In the second half of the 19th century, a cousin of Charles Darwin named Francis Galton seized upon Kelvin's suggestion as the basis for a number of unusual studies designed to measure the unmeasurable. For example, infuriated by the vagueness associated with a term like "beauty," Galton set out to create a "beauty map" of the British Isles, whereby each woman he came into contact with was classified as either "attractive, indifferent, or repellent."

London, he claimed, had the highest concentration of beautiful women, while Aberdeen was mathematically proven to be home to the ugliest.[16] Another study saw him measure listlessness by constructing a unified "Measure of Fidget," as Galton felt the "mutiny of constraint" epitomized in a fidget lent "numerical expression to the amount of boredom expressed by [an] audience." The more fidgeting, the higher the levels of boredom.[17] Even God wasn't safe from quantification, since Galton saw no reason that the "efficacy of prayer" (the rate at which prayers were answered versus ignored) should not be a "perfectly appropriate and legitimate subject of scientific inquiry."[18]

It is into this quantified space that Silicon Valley start-up Knack enters the picture. Founded by Israeli entrepreneur Guy Halfteck, Knack has a deceptively simple aim: to use a combination of gaming technology, machine-learning algorithms and the latest findings from behavioral science to come up with universal measures for terms like "quick-thinking," "perceptiveness," "empathy," "insightfulness," "spontaneity" and "creativity." By doing this, Halfteck says that he hopes to trigger a "fundamental change in the human capital space" that will seek to unlock an individual's previously untapped potential.

The basis for Knack's work is an insight that has been explored by psychologists for the last half century: that the way we play games can be used to predict how we behave in the real world. "Even though to your eye, your behavior in a game does not necessarily characterize your real-world behavior, it is highly likely that the way you play a game and another person plays that same game would reveal differences about personality and the way that your brain works," Halfteck says. "Your working memory, your strategic thinking, your risk-taking—these are all things which are manifested in how we game."

Knack's games currently include *Wasabi Waiter* and *Balloon Brigade*. Both are straightforward, pick-and-play affairs that nonetheless offer the player a number of different ways to compete. In *Wasabi Waiter*, for example, players take on the role of a waiter and chef as they take customers' orders and then prepare the dish that matches his or her facial expression. This expression might be "happy," "sad," "angry" or "any mood" in the event that the player is unsure. When a customer finishes eating, the player brings their plate back to the sink and starts the process again with someone new.

This may appear simple, but beneath the surface this is anything but. In a game, literally everything is measurable: each action, message, item and rule is composed of raw data. For every millisecond of play in *Wasabi Waiter* or *Balloon Brigade*, hundreds of data variables are gathered, processed and analyzed based on the decisions players make, the speed at which they do things, and the degree to which their game playing changes over time. What Halfteck perceives to be the accompanying behavioral qualities are then teased out using machine-learning tools and data-mining algorithms. "This is a very rich data stream we're collecting," Halfteck says. "It really allows us to get closer to the unique behavioral genome of a person."

There is, however, an innate danger in attempting to quantify the unquantifiable. When it comes to taking complex ideas and reducing these to measurable elements, the most famous critique came from evolutionary biologist Stephen Jay Gould. In his 1981 book, *The Mismeasure of Man*, Gould warned about the dangers of converting concepts such as "intelligence" into simplified measures such as IQ. Gould's concerns weren't just about the potential dangers of abstraction, but about the ways in which apparently objective truths can be used to back up human biases, rather than to expose genuine insights. In his own words, *The Mismeasure of Man* is an attack on "the abstraction of intelligence as a single entity, its location within the brain, its quantification as one number for each individual, and the use of these numbers to rank people in a single series of worthiness, invariably to find that oppressed and disadvantaged groups—races, classes, or sexes—are innately inferior and deserve their status."[19] Measurement and reductionism, of course, go hand in hand. Each time we begin to measure something, we lose whatever it is that the measurement tool is not designed to capture, or that the person measuring is not aware of the need to measure. As I will discuss in the coming chapters, technological attempts to create simple quantifiable measures for ideas like "creativity" and "love" meet with fierce opposition—largely because the concepts are far from simple.

But Halfteck disagrees that the beauty of terms like "empathy" and "insightfulness" is in their abstract amorphousness. "We are talking about an ever-expanding universe of things that are being measured," he says. "In the case of Knack, we're moving in a positive direction from a paradigm where people are being measured on a single dimension, to one in which people are measured in a polydimensional way on exponentially more aspects of their personality. It's not just about 'intelligence,' but rather the sum total of the human condition. It's far more nuanced than anything we've seen before."

Your Life According to Twitter

In November 2013, I wrote an article for *Fast Company* about two computer science researchers who had created an algorithm that used the information posted in your Twitter feed to generate customized user biographies.[20] It was a fascinating experiment in the subject of "topic extraction" and—based on the feedback and the number of hits the article received—I was not alone in feeling that way. Twitter, explained one of the study's authors, Jiwei Li, was the perfect tool for researchers. First, it encouraged its users to keep diaries in a publicly accessible medium. Second, the micro-blogging site's imposed 140-character limit for messages forced users to be concise: compressing complex life events into a few short sentences.

Since reading through a person's Twitter feed going back a number of years could prove prohibitively time consuming (particularly if you were doing this for a large number of people, as you might if you were the boss wanting to know more about the people who worked for you), the algorithm's job was to scan through this information and output it in a more accessible and easily readable format; pulling out only the relevant tidbits that might inform you about the major events in a person's life. The algorithm worked by analyzing the contents of each tweet and dividing these into "public" and "private" categories (say, your opinion on a major sporting event versus your own birthday) and then subdividing each of these categories again into "general" and "specific" headings. "General" events would typically be things like complaints about the commute to work or comments on your weekly yoga class, and would show their predictability by virtue of recurring over a long period of time. "Specific" events, on the other hand, would be life-changing events like a person's wedding, or the offer of a new job, and would frequently be the subject of a large amount of activity taking place over a short time span.

Unsurprisingly, it was this last "private-specific" event that was of most interest when it came to generating a mini-biography. As anyone who has ever read a popular biography will know, more space is typically given to the unique events in a person's life that are specific to them rather than a broader contextual look at where they fit within the wider society. Walter Isaacson's *Steve Jobs*, for example, focuses far more on Jobs's work creating the iPhone than it does on what he saw on his drive to and from his Palo Alto home each day.[21] But assuming that "public" or "general" observations are simply noise to be filtered out risks grossly simplifying a person's life by viewing them as existing in a context-free void. Wherever you are on the political spectrum, it is impossible to deny that public events like general elections or overseas conflicts have a bearing on the lives of the typical individual. The same is true of events that recur regularly, but are nonspecific in nature. The worker's commute to the office, or the inner-city family that has its power turned off or lives in a crime-ridden neighborhood, may not be "specific" but does as much, if not more, to explain their circumstances than which geo-tagged location they were in when they proposed to their partner.

I am not entirely blaming the two researchers for this. Their conception of the self is a neat one, which fits neatly within the Western value system that presents the individual as both a unique and fundamentally autonomous being. This idea forms not just the basis of the social sciences, but also politics, economics and the legal system. In politics, the self is the citizen who participates in democracy through voting and other political activities. In a market economy, the self is the optimizer of costs and benefits in such a way that they correspond with a person's best interests. In the legal system (which I explore in more detail in Chapter 3), the self is usually imagined as an agent who is responsible for his or her own behavior within society. What underlines all of these interpretations is the notion that at its root the self is a profoundly rational entity.

In today's digital world the conception of the self relies largely on the inferences of algorithms—comparing individual qualities against large data sets of "knowable" qualities to find correlations. These are, in a very real sense, formulas of the self. Some are extremely complex and depend on gathering as many data points as possible about particular people before coming up with conclusions. Others are about simplicity: as with abstract art, taking the broadest possible "shapes" that define a human being, *reductio ad absurdum*. One start-up named Hunch claims that with just five data points (in other words, a user answering five questions) it can answer practically any consumer preference question with 80 to 85 percent accuracy. YouAreWhatYouLike, meanwhile, offers to create detailed profiles of particular users by analyzing the common associations of their Facebook "likes" with a data set of "social dimensions." We are told, for instance, that users who "like" online art community deviantART.com are liberal in their political leanings, while those who "like" NASCAR tend toward the conservative. Other "likes" prove similarly predictive of personality types who might be "distrustful" or "reserved." A similar study carried out by University of

Cambridge researchers in 2013 suggested that algorithms provided with a dataset of 58,000 American Facebook users were able to accurately predict qualities and traits, including race, age, IQ, sexual preference, personality, substance use and political views—all based upon "likes."[22] Another service—TweetPsych—claims to use algorithms to score a person's emotional and intellectual quotients based upon the topics they choose to tweet about, including learning, money, emotions and anxiety. Yet more studies have been shown to be able to deduce gender, sexual orientation, political preference, religion and race with a greater than 75 percent level of accuracy. A 2010 investigation by psychologist Tal Yarkoni of the University of Colorado at Boulder analyzed the words in 695 blogs and compared these to the personalities of their owners as revealed through personality tests. Yarkoni suggested that neurotic bloggers are more likely to use words like "awful" and "lazy," conscientious ones overuse "completed," and generally agreeable ones fall back on describing things as "wonderful."[23]

It is the gulf between the idea of the autonomous individual and the algorithmic tendency to view the individual as one categorizable node in an aggregate mass that can result in The Formula's equivalent of a crisis of self. Writing in 2012, a Facebook user commented on the new Timeline feature being rolled out in the social network's user interface at the time. Unlike the previous Facebook user interface, the Timeline had the narrativizing effect of proceduring history into a series of events (jobs, relationships) with unknown categories marked by blank spaces so that the implication was that the user should continue adding materials to their Timeline—retroactively tracing their own personal narrative back until they reach the category "born." "I feel betrayed by . . . an interface that appears to give so many choices on the surface, while limiting almost every bit of our creative endeavor to the predefined and prepackaged boxes and categories within which we're supposed to find a place," the user noted.

> It hurts us all, in different, small ways. Sure, I feel fine clicking the "female" category, but I know at least two dozen friends who wouldn't be able to choose a box. I'm supposed to declare a "hometown" but I've not had one for more than twenty years, so that's not useful at all. I now have to mark places on a map, or accept the default map that appeared on my profile just this afternoon . . . What if I don't want to be defined by time or any other moment that Facebook has determined is "relevant" in my life?[24]

Big Brother, Sort Of

A number of cultural critics have commented upon the large number of ways in which bureaucratic measures have intensified under neoliberalism—despite its presentation as being profoundly and fundamentally antibureaucratic by nature.

Such critiques can certainly be applied to those high-tech companies in thrall to The Formula. One example is the high-tech start-up CourseSmart, which allows teachers to surveil their students even when they are away from the classroom. Much like e-book analytics that can be fed back to publishers (something that I will describe later on in this book), Course-Smart uses algorithms to track whether students are skipping pages in their textbooks, not highlighting significant passages, hardly bothering to take notes, or even failing to study at all. In April 2013, CourseSmart was the subject of an article in the *New York Times*, under the headline "Teacher Knows If You've Done the E-Reading." The story related the plight of a teacher who tracked 70 students in three of his classes. Despite one student regularly scoring high marks in mini-tests, Course-Smart alerted the teacher that his student was doing all of their studying the night before tests, as opposed to taking a long-haul approach to learning. The article quoted the university's school of business dean as describing the service as "Big Brother, sort of, but with a good intent."[25] According to the story:

> Students do not see their engagement indexes (Course-Smart's proprietary analytics tool) unless a professor shows them, but they know the books are watching them. For a few, merely hearing the number is a shock. Charles Tejeda got a C on the last quiz, but the real revelation that he is struggling was a low CourseSmart index.
>
> "They caught me," said Mr. Tejeda, 43. He has two jobs and three children, and can study only late at night. "Maybe I need to focus more," he said.

On the surface, CourseSmart offers considerably more free-dom than the kind of factory model "industrial schooling" that rose to prominence with the Industrial Revolution, pitting warden-teachers against prisoner-students. Students are less class-room-bound and are afforded opportunities to study on their own. (Or, at least, ostensibly on their own.) However, such tools actually represent a more continuous form of control system based on increasingly abstract entities like "engagement." After all, as the *New York Times* story demonstrates, a person could receive a "satisfactory" C grade, only to fail a class on engagement.

A parallel to CourseSmart is the kind of deep data analytics Google uses to track its own workforce. Like many high-tech businesses, Google models itself as a libertarian utopia: the type of company where employees used to be allowed one extra day per week to pursue their own lines of inquiry, and are as likely to spend their time ascending Google's indoor rock-climbing wall or having free food served up to them by a former Grateful Dead chef as they are to be coding. However, as Steven Levy points out in *In the Plex*, his 2011 study of Google, the search leviathan's apparent loopiness is "the crazy-like-a-fox variety and not the kind calling for straightjackets."[26] Despite Google's widely publicized quirks, its irreverent touches are data-driven to a fault. "At times Google's largesse can sound excessive," notes an article in *Slate*. "Yet it would be a mistake to conclude that Google doles out such perks just to be nice. [The company] rigorously monitors a slew of data about how employees respond to benefits, and . . . rarely throws money away."[27]

For instance, there is a dedicated team within Google called the People Analytics group, whose job is to quantify the "happiness" of employees working for the company. This is done using "Googlegeist," a scientifically constructed employee survey, which is then mined for insights using state-of-the-art proprietary algorithms. An example of what the People Analytics team does occurred several years ago, when Google noticed that a larger-than-normal number of female employees were leaving the company. Drilling down with data-mining tools, the People Analytics group discovered that this wasn't so much a "woman" problem as it was a "mother" problem: women who had recently given birth were twice as likely to leave Google as its average departure rate. The most cost-effective answer, it was concluded, was to increase maternity leave from the standard 12 weeks of paid absence to a full five months. Once the problem had been identified and acted upon, the company's attrition rate for new mothers dropped by 50 percent.

Similar data-driven insights are used to answer a plethora of other questions. Just how often, for example, should employees be reminded to contribute to pension plans, and what tone should best be used when addressing them? Do successful middle managers possess certain skills in common, and could these be taught to less successful managers? And what is the best way to maximize happiness and, thus, efficiency in staff? A salary increase? Cash bonus? Stock options? More time off?

For all its hiding behind the image of soft "servant leadership" the real thing an entity such as Google's People Analytics group returns to prominence is the concept of "Taylorism." Created in the early 20th century by engineer Frederick Taylor, the ideas behind Taylorism were outlined in a 1911 book called *The Principles of Scientific Management*.[28] At the center of Taylor's beliefs was the idea that the goal of human labor and thought should be increased efficiency; that technical calculation is always superior to human judgment; that subjectivity represents a dumbing-down of clear-thinking objectivity; and that whatever is unable to be quantified either does not exist or has no value. "It is," he argued, "only through enforced standardization of methods, enforced adoption of the best implements and working conditions, and enforced cooperation that . . . faster work can be assured."

Work Faster and Happier

Of course, it's not just about faster work. As the quantification of "happiness" and "engagement" demonstrate, it is no longer enough to simply be an effective laborer. A person must also be an *affective* laborer, offering "service with a smile." In this way it is necessary to ask the degree to which The Formula is genuinely improving working conditions, or whether it is simply (to quote cultural critic Paul Virilio) transforming workers into unwitting participants in a Marxist state pageant, "miming the joys [of] being liberated"—with full knowledge that anything other than full enthusiasm will be noted down against their CourseSmart-style engagement index and unearthed by an algorithm in time for a future job interview.

Here it is worth turning once more to the work of Alvin Toffler, whose concept of "demassification" laid out many of the principles described in this chapter. In *The Third Wave*, Toffler questions why it is that everyone should be asked to start work at 9 A.M. and finish at 5 P.M. each day. By changing this massified approach to a more individual one, centered around the self, Toffler argues that both employers and employees would experience benefits. The former group could use insights about the times their paid employees are at their most productive to maximize efficiency. Employees, meanwhile, could arrange their working hours around their other nonwork duties, or simply their natural biological rhythms (which we now know can be ascertained through wearable sensors). Of course, what sounded a utopian formula to Toffler now exists as the commonplace "flexitime" approach to employment, in which many companies have laid off their

permanent workforce in favor of a free-floating pool of part-
time, fixed-term and freelance workers lacking in benefits and
job security. In Becky Hogge's *Barefoot into Cyberspace*, the
author relates this directly to the dream of the techno-
solutionists, noting how "the eighties personal computer
gurus are . . . the same folk who went around major corpora-
tions advising them on ways to decouple their fortunes from
those of their employees, ushering in the era of precarious
employment that is my generation's norm."[29] In the gamified
collapse of work into play and play into work, concepts like
performance-based pay (presented as another level of personalization) mean that even those jobs that do not immediately
lend themselves to increased speed and efficiency can be subjected to The Formula.

This neo-Taylorist dynamic becomes more apparent the further you look down the high-tech food chain. In Amazon's
warehouses, for example, product pickers (known as "fulfill-
ment associates") are issued handheld computers that transmit
instructions to reveal where individual products can be picked
up or dropped off. Because of the size of Amazon's warehouses, a routing algorithm is used to work out the shortest
possible journey from point to point. That is not all the handheld computers do, however. They also collect a constant, real-
time stream of data that monitors how fast employees walk and
complete individual orders, thus quantifying their productivity.[30] Like the bomb-loaded bus in the movie *Speed*, workers
must maintain a certain minimum speed, or else see their jobs
go up in smoke. As with Henry Ford's assembly lines, so too
here does machinery determine the pace of work. A warehouse
manager at Amazon has been quoted as describing workers as

"sort of like a robot, but in human form."[31] An article for *Fast Company* paints a depressingly Orwellian picture:

> An Amazon fulfillment associate might have to walk as far as 15 miles in a single shift, endlessly looping back and forth between shelves in a warehouse the size of nine soccer fields. They do this in complete silence, except for the sound of their feet. The atmosphere is so quiet that workers can be fired for even talking to one another. And all the while, cardboard cutouts of happy Amazon workers look on, cartoon speech bubbles frozen above their heads: "This is the best job I ever had!"[32]

Similar reports can be seen elsewhere. In Tesco warehouses in the UK, workers are made to wear arm-mounted electronic terminals so that managers can grade them on how hard they are working. Employees are allocated a certain amount of time to collect an order from a warehouse and then complete it. If they meet the target, they are awarded a 100 percent score, rising to 200 percent if the task is completed in double time. Conversely, scores fall dramatically in situations where the task takes longer than expected.[33]

Decimated-Reality Aggregators

Speaking in October 1944, during the rebuilding of the House of Commons, which had sustained heavy bombing damage during the Battle of Britain, former British prime minister Winston Churchill observed, "We shape our buildings; thereafter they shape us."[34] A similar sentiment might be said in the age of The Formula, in which users shape their online profiles, and from that point forward their online profiles begin to shape them—both in terms of what we see and, perhaps more crucially, what we don't.

Writing about a start-up called Nara, in the middle of 2013, I coined the phrase "decimated reality aggregators" to describe what the company was trying to do.[35] Starting out as a restaurant recommender system by connecting together thousands of restaurants around the world, Nara's ultimate goal was to become the recommender system for your life: drawing on what it knew about you from the restaurants you ate in, to suggest everything from hotels to clothes. Nara even incorporated the idea of upward mobility into its algorithm. Say, for example, you wanted to be a wine connoisseur two years down the line, but currently had no idea how to tell your Chardonnay from your Chianti. Plotting a path through a mass of aggregated user data, Nara could subtly poke and prod you to make sure that you ended up at a particular end point after a certain amount of time. If you trained Nara's algorithms to recognize what you wanted, Nara's algorithms could then train you to fit a certain desired mold. In this way, "decimated reality" was a way of getting away from the informational overload of the Internet. Users wouldn't see more options than they could handle—they would see only what was deemed relevant.

"The Internet has evolved into a transactional machine where we give our eyeballs and clicks, and the machine gives us back advertising and clutter," says Nathan Wilson, chief technology officer at Nara Logics Inc. "I'm interested in trying to subvert all of that; removing the clutter and noise to create a more efficient way to help users gain access to things."

The problem, of course, is that in order to save you time by removing the "clutter" of the online world, Nara's algorithms must make constant decisions on behalf of the user about what it is that they should and should not see.

This effect is often called the "filter bubble." In his book of the same title, Eli Pariser notes how two different users searching for the same thing using Google will receive very different sets of results.[36] A liberal who types "BP" into his browser might get information about the April 2010 oil spill in the Gulf of Mexico, while a conservative typing the same two letters is more likely to receive investment information about the oil company. Similarly, algorithms are more likely to respond to a female search for "wagner" by directing the user toward sites about the composer "Richard Wagner," while a male is taken as meaning "Wagner USA" paint supplies. As such, what is presented by search algorithms is not a formula designed to give ideologically untampered answers but precisely the opposite: search results that flatter our personal mythologies by reinforcing what we already "know" about particular issues, while also downgrading in importance the concerns that do not marry up to our existing worldview.

Despite the apparent freedom of such an innovation (surely, personalization equals good?), it is not difficult to see the all-too apparent downside. Unlike the libertarian technologist's pipe dream of a world that is free, flat and open to all voices, a key component of code and algorithmic culture is software's task of sorting, classifying and creating hierarchies. Since so much of the revenue of companies like Google depends on the cognitive capital generated by users, this "software sorting" immediately does away with the idea that there is no such thing as a digital caste system. As with the "filter bubble," it can be difficult to tell whether the endless distinctions made regarding geo-demographic profiles are helpful examples of mass customization or exclusionary examples of coded discrimination. Philosopher Félix Guattari imagined the city in which a person was free to leave their apartment, their street or their neighborhood thanks to an electronic security card that raised barriers at each intersection. However, while this card represents freedom, it also represents repression, since the technology that opens doors for us might just as easily keep them shut. Much the same might be said of the algorithm, which is directly and automatically responsible for providing social and geographical access to a number of goods, services and opportunities for individuals.

The idea that a certain class of user can be willfully inconvenienced in favor of another more desirable class is a key part of market segmentation. At various times, UK supermarkets have investigated the possibility of charging a premium for food shopping at peak shopping hours, in an effort to deter the "cash-poor but time-rich" customers from negatively impacting upon the shopping experience of the more desirable "cash-rich but time-poor" professionals.[37] Many airlines offer premium schemes that allow valuable business travelers to pay an extra surcharge to bypass certain border controls. Sorting processes divide passengers into groups of those enrolled on the scheme and those that are not. In the case of the former premium group, passengers are offered parking spaces close to the airport terminal and then allowed to pass through to dedicated members' lounges with speed. In the case of the latter "cash-poor but time-rich" group, assigned parking spaces are typically a long distance from the airport terminal, and passengers are excluded from VIP lounges and forced to endure lengthy "check-in" times and security queues. A similar brand of thinking is behind schemes set up in cities such as Los Angeles, San Diego, Toronto, Melbourne and Tokyo, in which privately funded highways are built for use by affluent motorists, who access them by paying a premium fee. To keep these highways exclusive, algorithms are used to estimate the exact level of price per journey that is likely to deter enough drivers so as to guarantee free-flowing traffic, regardless of how bad congestion might be on the surrounding public highway system.[38]

Squelching the Scavengers

The digital world is not immune to these practices. In the past, Internet networking provider Cisco has referred to its less-than-premium consumers as a "scavenger class" for whom the company provides "less-than-best-effort services" to certain applications. At times of Internet congestion and slow download times, traffic can be (in Cisco's words) "squelched to virtually nothing" for scavenger-class users, while more valued business users are provided with access to Cisco's answer to free-moving, privately funded highways. Similar distinctions might be made elsewhere. A marketing brochure from communications company the Avaya Corporation similarly promises that its bespoke systems allow for algorithms to compare the numbers of individual call-center callers to a database, and then route calls through to agents as high-priority if the caller is in the top 5 percent of customers. In this scenario, when the agent picks up the call, they hear a whispered announcement that the caller they are speaking with is "Top 5." Much the same technology could be put into practice in situations where algorithms are used to filter the personalities of individual callers. Callers might receive service according to personality types, so that more lucrative customers with an increased likelihood of losing patience with a service quicker could be ranked above personality types likely to procrastinate and ask lots of questions, while being unlikely to spend large amounts of money with a company in the future.

Perhaps the most granular example of this "human software sorting" can be seen with the algorithm-driven idea of "differential pricing." This is something a number of online stores, including Amazon, have already experimented with. Without the added gloss of marketing speak, differential pricing means that certain users can be charged more than others for the same product. In the abstract, the algorithms used for this are no different from those that predict that, since a person bought the *Harry Potter* and *Twilight* books, they might also enjoy the *Hunger Games* trilogy. It is presented as another level of personalization, in line with the website that remembers your name (and, more importantly, your credit-card details). In September 2012, Google was granted a patent for "Dynamic Pricing on Electronic Content," which allows it to change the listed price of online materials, such as video and audio recordings, e-books and computer games, based upon whether its algorithms determine a user is more or less likely to purchase a particular item. The filed patent was accompanied by illustrations gleefully suggesting that certain users could be convinced to pay up to four times what others are charged for the exact same digital file.[39] In other words, if

Google's algorithms "know" that you are susceptible to tween-fodder like *Harry Potter* and *Twilight*, based on what you have searched for in the past, it can ensure that you pay through the nose for *The Hunger Games*—while also enticing the person who has only ever demonstrated the slightest of interests in teenage wizards and sparkly vampires to buy the product by lowering the price to tempt them.

Recent years have also seen a rise in so-called emotion sniffing algorithms, designed to predict a user's emotional state based on their tone of voice, facial expression—or even browsing history. A study carried out by Microsoft Research analyzed people's phone records, app usage and current location, and then used these metrics to predict their mood. According to Microsoft, the algorithm's daily mood estimates start at 66 percent accuracy and gradually increase to 93 percent after a two-month training period.[40] Since mood significantly influences consumer preferences, information like this could prove invaluable to marketers.[41] Let's say, for example, that your computer or smartphone determines that you're likely to be feeling particularly vulnerable at any given moment. In such a scenario it might be possible to surreptitiously raise the price of products you are likely to be interested in since an algorithm designed to "sniff" your mood has determined that you're statistically more likely to be susceptible to a purchase in this state.

Free-market idealists might argue in favor of such an approach. In the same way that an upmarket restaurant charges more for the exact same bottle of beer that a person could buy for half that price somewhere else, so concepts like differential pricing could be used to ensure that a person pays the exact price he or she is willing to spend on a product—albeit with more scientific precision. However, while this is undoubtedly true, the more accurate analogy may be the restaurant whose waiters rifle through your belongings for previous receipts before deciding what they think you ought to be charged for your food and drink. Differential pricing could be used to even the playing field (everyone pays 1/250 of their weekly salary for a beer, for example). More likely, however, it could be used to do just the opposite: to raise the price of particular goods with the stated aim of marginalizing or driving away those less lucrative—and therefore less desirable—users. A high-fashion brand might, for instance, want to avoid becoming associated with a supposed "undesirable" consumer base, as happened to Burberry in the early 2000s. Since it cannot outright bar particular groups from purchasing its products, the company could pay to have its products disappear from the algorithmic recommendations given to those who earned under $50,000 per year or—in the event that its products were specifically searched for—priced out of their range.

The real problem with this is the invisibility of the process. In the same way that all we see are the end results when algorithms select the personalized banners that appear on websites we browse, or determine the suggested films recommended to us on Netflix, so with differential pricing are customers not informed that they are being asked to pay more money than their next-door neighbor. After all, who would continue shopping if this were the case? As Joseph Turow, a professor at the University of Pennsylvania's Annenberg School for Communication and frequent writer about all things marketing, has pointed out in an article that appeared in the *New York Times*: "The flow of data about us is so surreptitious and so complex that we won't even know when price discrimination starts. We'll just get different prices, different news, different entertainment."[42]

The Discrimination Formula?

A number of Internet theorists have argued that in the digital world, previous classifications used for discrimination (including race, gender or sexuality) will fall away—if they haven't already. Alvin Toffler's *Third Wave* identifies a number of individuals and groups subtly or openly discriminated against during the last centuries and argues that this marginalization is the product of Second Wave societies. According to Toffler, in the unbundled, personality-driven Third Wave society such discriminatory practices will slink off into the digital ether. This is a popular utopian view. In Mark Hansen's essay "Digitizing the Racialized Body, or The Politics of Common Impropriety," the author builds on this point by suggesting that the web makes possible an unprecedented number of opportunities for ethical encounters between people of different races, since race as a visual signifier is rendered invisible.[43] The mistake made by both Toffler and Hansen is assuming that discrimination is always collectivist in nature. The Formula suggests that this is far from the case. In an age in which power no longer has to be embodied within a set structure and can be both codified and free-floating, discrimination is able to become even more granular, with categories such as race and gender made increasingly nonstatic and fluid. As can be seen with Internet-dating profiles, which I describe in more detail in Chapter 2, in order for algorithmic sorting to take

place, individuals must first be subjected to the process of segmentation, where they are divided up into their individual components for reasons of analytics. Instead of being "individuals,"
they are turned into "dividuals."

The concept of "dividuals" is not mine. The French philosopher Gilles Deleuze coined this phrase to describe physically
embodied human beings who are nonetheless endlessly divided and reduced to data representations using tools such as
algorithms. The Formula, Deleuze and his coauthor Félix
Guattari argue in *A Thousand Plateaus*, has turned man into a
"segmentary animal."

> We are segmented in a *binary* fashion, following the great
> major dualist oppositions: social classes, but also men-
> women, adults-children, and so on. We are segmented in a
> *circular* fashion, in ever larger circles, ever wider disks or
> coronas, like Joyce's "letter": my affairs, my neighborhood's
> affairs, my city's, my country's, the world's . . . We are seg-
> mented in a *linear* fashion, along a straight line or a num-
> ber of straight lines, of which each segment represents an
> episode or "proceeding": as soon as we finish one proceed-
> ing we begin another, forever proceduring or procedured,
> in the family, in the school, in the army, on the job.[44]

Deleuze expanded on this idea later in his life by discussing
what he called the "society of control."[45] Echoing Toffler's
Third Wave thesis that we have progressed from a disciplinary
society based on the production of physical goods, to an economy founded on information and financialization, Deleuze
examined how the structures of power and control have

changed. In previous societies, he argues that control occurred within specific linear sites, such as the school, the workplace or the family home. Each one of these came with its own unique set of rules, which applied only to that site. Between spaces, people were relatively unmonitored and still had space for uncontrolled life to occur. This changes in a control society, in which there is continuous regulation, although this is less obvious than in a disciplinary society. Since power is everywhere it also appears to be nowhere. Rather than being forced to fit into preexisting molds, Deleuze argues that we are encased in an enclosure that transforms from moment to moment—like a giant sieve whose mesh strains and bulges from point to point.

This seems to do a good job of summing up the new algorithmic identity. In the digital age, everyone can have a formula of his or her own. Companies like Quantcast and Google get no benefit at all from everyone acting in the same way, since this allows for no market segmentation to occur. It is for this reason that articles like Steven Poole's May 2013 cover story for the *New Statesman*, "The Digital Panopticon," invoke the wrong metaphor when it comes to big data and algorithmic sorting.[46]

The panopticon, for those unfamiliar with it, was a prison designed by English philosopher and social theorist Jeremy Bentham in the late 18th century. Circular in design and with a large watchtower in the center, the theory behind the panopticon was that prisoners would behave as if they were being watched at all times—with the mere presence of the watchtower being as effective as iron bars in regulating behavior and ensuring that everyone acted the same way. (A similar idea is behind today's open-plan offices.)

As former *Wired* editor Chris Anderson argues in *The Long Tail*, modern commerce depends upon market segmentation.[47] Unless you're selling soap, aiming a product at a homogeneous mass audience is a waste of time. Instead, vendors and marketers increasingly focus on niche audiences. For niches to work, it is important to companies that they know our eccentricities, so that they can figure out which tiny interest group we belong to. In this sense, an authoritarian device like the panopticon—which makes everyone act the same way—is counterproductive. In algorithmic sorting, audiences know they are being surveilled; they just don't care. The apparatus of capture (how companies recognize us) and the apparatus of freedom (buy the products that best sum you up as an individual) are entwined so totally as to be almost inseparable. As French philosopher Jacques Ellul argued in *The Technological Society*, the citizens of the future (he was writing in the early 1960s) would have everything their heart desired, except for their freedom.[48] This chilling concept is one that was more recently expanded upon by media historian Fred Turner:

> If the workers of the industrial factory found themselves laboring in an iron cage, the workers of many of today's post-industrial information firms often find themselves inhabiting a velvet gold mine . . . a workplace in which the pursuit of self-fulfillment, reputation, and community identity, of interpersonal relationships and intellectual pleasure, help to drive the production of new media goods.[49]

In her work on the digital identity, MIT psychoanalyst Sherry Turkle talks about our different "selves" as windows.

"Windows have become a powerful metaphor for thinking about the self as a multiple, distributed system," she wrote in her 1995 book *Life on Screen*, a magnum opus that landed her on the cover of *Wired* magazine. "The self is no longer simply playing different roles in different settings at different times. The life practice of windows is that of a [de-centered] self that exists in many worlds, that plays many roles at the same time."[50]

This view of the self as a "multiple, distributed system" (or else a slightly clunky Microsoft operating system) was meant as empowering. The de-centered, windowed self means that the woman who wakes up in bed next to her husband can walk downstairs, close the "wife" window and open the "mother" one in order to make breakfast for her daughter, before hopping in the car and driving to work, where she will once again clear her existing windows and open a new one titled "lawyer" or "doctor." This is, of course, the thing about windows: they can be opened and closed at will.

What Fred Turner describes, on the other hand, is a world in which multiple subjectivities exist, but these subjectivities constantly crash into one another. Unlike the "windowed self" or the "segmentary animal" who fills different roles at school, in the workplace and at the home, where The Formula is involved these rules are not isolated to one location but affect one another in intricate, granular and often invisible ways.

Keeping Up Appearances

It might not even matter whether specific pieces of data shaping our identity are "true" or not. Whether we are physically male or female, or else consider ourselves to be male or female, ultimately what will determine how we are treated online is the conclusions reached by algorithms. A *New York Times* article from April 2013 related the story of an unnamed friend of the writer's who received, by mail, a flyer advertising a seminar for patients suffering from multiple sclerosis, hosted by two drug companies, Pfizer and EMD Serono. Spam e-mails and their physical counterparts are, of course, nothing new. What was alarming about this situation, however, was not the existence of the spam message, but the targeting of it. The recipient was not an MS sufferer, although she had spent some time the previous year looking up the disease on a number of consumer health sites. As the arrival of the flyer proved, somewhere her name and contact details had been added to a database of MS sufferers in Manhattan. The ramifications of this were potentially vast. "Could [my friend], for instance, someday be denied life insurance on the basis of that profile?" the author asks. "She wanted to track down the source of the data, correct her profile and, if possible, prevent further dissemination of the information. But she didn't know which company had collected and shared the data in the first place, so she didn't know how to have her entry removed from the original marketing list."[51]

John Cheney-Lippold, the scholar who coined the term "new algorithmic identity," says that the top-performing male students in his classroom are regularly characterized as female online. "Who is to say that they're not female in that case?" he asks rhetorically. In a world in which terms like "male" and "female" are simply placeholders for specific types of behavior, traditional distinctions of discrimination break down. With this in mind, who is to say what "gender" or "racial" discrimination will look like by, for example, the year 2040? If gender discrimination is not based on anything physical but rather on the inferences of algorithms, can a man be discriminated against in terms of the barring of certain services because he skews statistically female in his clicks? What does it mean for the future of racial politics if a young white male growing up in a working-class environment in inner-city Detroit is classified "blacker" than an older, educated African-American female living in Cambridge, Massachusetts? Could a person be turned down for a job on the basis that it is a role black males statistically do worse in (whatever that might be), even if the individual in question is physically a white female?

These types of discriminatory behavior could prove challenging to break, particularly if they are largely invisible and in most cases users will never know how they have been categorized. Unlike the shared history that was drawn on to help bring about the civil rights or women's lib movements, algorithmically generated consumer categories have no cultural background to draw upon. What would have happened in the case of Rosa Parks's December 1955 protest—which garnered the support of the African-American community at large—had she not been discriminated against purely on the basis of her skin color, but on several thousand uniquely weighted variables based upon age, location, race and search term history? There is racism, ageism and sexism, but is there an "ism" for every possible means of demographic and psychographic exclusion? Unlike previous discriminatory apparatuses, today's categories of differentiation may be multiply cross-referenced to the point where it even becomes difficult to single out the single overriding factor that leads to a person being denied credit, or enables them to proactively alter their perceived desirability.

It is also worth noting that gender and race do not exist as stable concepts but rather in a state of constant flux. Like the concept of a "character" that a person builds but never finishes, so too are specific categories like maleness not simply inferred by algorithms and then established from that point on, but instead have to be reinforced on a constant basis.[52] A user may be considered male, but in the event that they then begin searching for more "female" subjects, they will be reclassified. The man who regularly buys airline tickets might be recognized as increasingly female, while a female with a keen interest in sports or world news becomes statistically male.

Categorizing additionally has the ability to move beyond what we might traditionally think of as categories. If concepts like "creativity" and "perceptiveness" are successfully quantified and linked to consistent types of behavior, these might take on as much importance as gender or race. It was this world that Deleuze was addressing when he predicted that increasingly our credit cards and social security numbers will become more significant identity markers than the color of our skin or the place we went to school. "A man is no longer a man confined," he wrote, "but a man in debt."

CHAPTER 2

The Match & the Spark

Each summer, thousands of freshly qualified doctors graduate from medical school in the United States. The next step of their training involves being paired up with a teaching hospital where they can continue to learn their craft; a process that is referred to as residency.

Deciding which doctors go to which hospitals involves a two-way matching process in which both doctors and hospitals have their list of preferences, and the task is to match both parties up in such a way that everyone is pleased with the decision.

Of course, this is easier said than done. Among both people and hospitals, some are more popular and in higher demand than others. A hospital might be geographically preferable, for example: perhaps situated in a big city, or in an especially scenic location. It might be preferable based on its overall reputation, or because of a particular member of staff who is held in particularly high regard within the medical teaching community.

Think about how difficult it could be to decide, based on consensus, where to go on a family holiday. Now imagine that each location around the world can only be visited by one family at a time, but that all families have to still go on holiday during the same week. Now imagine that those holiday destinations also have their own list about which family they want to welcome.

In 1962, two American economists named David Gale and Lloyd Shapley set out to devise an algorithmic solution to just this conundrum. What they came up with was called The Stable Marriage Problem—also known as "The Match."[1]

To explain The Match, picture a remote island, cut off from the rest of civilization. On this island, there live an even number of men and women of marriageable age. The problem asks that all of these people are paired up in what we might consider a stable relationship. To explain what is meant by "stable" first allow me to explain what counts as an "unstable" marriage. Suppose that two of the marriageable men on this island are named James and Rob, while two of the marriageable women are named Ruth and Alice. James is married to Ruth, although he secretly prefers Alice. Alice is married to Rob, but she secretly prefers James. Both, in other words, are already married to other people, but would be better off if they were matched together. It is not beyond the limits of our imagination to suppose that one day James and Alice will run off together—hence the lack of stability. The goal of the algorithm is to come up with a solution in which everyone is matched up in such a way that no couples exist that would rather be paired with a person other than with their respective partners.

There are a few rules to consider before any actual matching takes place. At the start of the problem, every marriageable man and woman on the island is asked to compile a list of preferences, in which they rank every member of the opposite sex in the order in which they most appeal. James might list Alice first, then Ruth, then Ruth's friend Charlotte, and so on. Only men may propose to women, although women have the right to not only refuse to marry a particular man if they deem him a bad match, but also to accept proposals on a tentative basis, keeping their options open in case someone better comes along. The marriage proposal process works in rounds, which I will describe as "days" in order for things to be kept simple.

On the morning of the first day, every man proposes to his first choice of wife. Certain women will be in the fortunate position of having received multiple proposals, while others will have received none. On the afternoon of the first day, each woman rejects all suitors except for her current best available option, who she tentatively agrees to marry, knowing that she can ditch him later on. (This is referred to as a "deferred acceptance.") Come dawn of the second day and those men who remain single propose to their next best choice. That afternoon, the women who accepted marriage on day one have the chance to trade up, if the man who proposed to them on day two is, in their view, preferable to the person they are engaged to. This process continues for days three, four, five, et cetera, until all couples are matched in stable relationships. At this point the algorithm terminates.

Here's an example:

Name	First Choice	Second Choice	Third Choice	Fourth Choice
Alice	Tim	James	Rob	Rajiv
Ruth	James	Rajiv	Tim	Rob
Charlotte	Rob	Rajiv	Tim	James
Bridgette	Rajiv	Rob	James	Tim

James	Alice	Bridgette	Charlotte	Ruth
Rob	Alice	Ruth	Charlotte	Bridgette
Tim	Ruth	Bridgette	Charlotte	Alice
Rajiv	Charlotte	Alice	Ruth	Bridgette

On day one, each man proposes to his top choice. James and Rob propose to Alice, Tim proposes to Ruth, and Rajiv proposes to Charlotte. Of Alice's two proposals she prefers James to Rob and so accepts his offer; knowing that she might well do better later on. Ruth, meanwhile, accepts Tim's proposal, while Charlotte accepts Rajiv's. On day two, Rob—rejected by Alice on the first day—proposes to Ruth. Rob, however, is Ruth's fourth choice and she remains engaged to Tim. On day three, Rob tries again and asks Charlotte to marry him. Rob is Charlotte's first choice and so she ditches Rajiv and becomes engaged to Rob. On day four, the newly single Rajiv asks Alice to marry him, although she elects to stay with James. On day five, Rajiv then asks Ruth to marry him, and Ruth breaks it off with Tim and becomes engaged to Rajiv. On day six, Tim proposes to Bridgette who, while he remains her fourth and last choice of match, has no other proposals and so agrees to marry him. The final couples are therefore as follows:

James and Alice	Rob and Charlotte	Rajiv and Ruth	Tim and Bridgette

There are several neat attributes to this particular algorithm. Among its most impressive qualities is the fact that not only does it manage to find a partner for everyone, but it does this with maximum efficiency. It is, of course, impossible in all but the most unlikely of cases that everyone will receive their first choice of partner, and this effect is only amplified as the numbers increase. Four boys and four girls may well receive their first choice, but would 40 boys and 40 girls? Or 400?

This algorithm, it should be noted, favors whoever it is that is doing the proposing (in this case the men). Were it to work the other way around from this male-optimal scenario (with the women asking the men to marry them, and the men doing the rejecting or the accepting) the algorithm for solving this particular problem would have terminated after one day, with the couples arranged as follows:

Alice and Tim	Ruth and James	Charlotte and Rob	Bridgette and Rajiv

For all its good qualities, however, there are a few conceptual problems with this problem, which follow as such. For one thing, The Match imagines that all of the men and women on the island are heterosexual, and therefore wish to be paired with members of the opposite sex. This statistically unlikely event is done only for the sake of mathematical simplicity. A scenario where men want to marry men, women want to marry women, or everyone wants to marry everyone, does not necessarily yield the same stable outcome as the relatively straightforward matching I described.

Another issue is that the Stable Marriage Problem presumes that all people of marriageable age do, in fact, wish to get married, and that the location in which they live is so remote that there is no chance whatsoever that any residents could marry someone from outside its boundaries. Yet another problem is that the algorithm assumes marriages only fail in the event of a disruptive third party, who is a better match for one part of a particular couple. While it is undeniably true that a certain percentage of marriages do end for this exact reason, it is by no means a universal, statistically robust rule. Some couples simply find themselves incompatible and conclude that they would be happier living on their own than they would with one another. To return to our original example, perhaps Alice falls head over heels in love with James only to realize after living with him for several months that her idea of a perfect evening is to go out dancing, while James would rather stay in and play computer games. Or maybe James chews with his mouth open, and bad manners are a deal-breaker for Alice.

The part of The Match that causes me the most problems, however, is the piece of information we are asked to take as writ before the algorithm even gets going: namely the idea that each of the men and women addressed by the puzzle is able to compile a list in which they rank, without error, everyone of the opposite gender in the order that they would most happily be married to them. Of all the assumptions made by David Gale and Lloyd Shapley, this seems the most grievous.

I do, of course, write these words with tongue firmly planted in cheek. As noted, the Stable Marriage Problem invokes romantic marriage as metaphor only, being designed for the purpose of matching medical students with hospitals. For this task it is largely suitable—since issues of one party chewing with their mouth open, or ogling a third party, are unlikely to result in separation. But by pointing out these conceptual difficulties, I do make a serious point: that outside of mathematical puzzles human beings have a nasty habit of behaving unpredictably.

And particularly when love is involved.

Madness in Love, Reason in Madness

In 2006, the pop statistician Garth Sundem was asked by the *New York Times* to create a formula to predict the breakup rate of celebrity marriages.[2] Sundem was well known for what he calls his "math in everyday life" equations. He'd appeared previously on the BBC and *Good Morning America*, and his formulas had proven so popular that he had published a book of them, *Geek Logik*, which contained mathematical answers to determine everything from whether a person should get a tattoo to how many beers they might want to take on their next work picnic. When the *New York Times* contacted him, Sundem says that he did what he always does when starting work on a new equation: he sat down at his desk and began thinking of criteria to test against the data on celebrity divorce rate. Did, for instance, the hair color of the couple in question make them more or less likely to divorce? How about the proximity of their home in relation to the Hollywood sign? In most cases, the answer was a resounding no. But data-mining did eventually produce results—as it always does. For example, Sundem discovered that the number of Google hits for a starlet that showed her in skimpy clothing positively correlates with the duration of her marriage. So too did the combined numbers of the couple's previous marriages.

Here is the formula he eventually came up with:

$$2\left(\frac{30-S^2}{P+5}\right)\left(\frac{10A_b+10A_g}{(A_b-A_g)^2+100}\right)\sqrt{\frac{75(10+G_b)}{(G_b+G_g)(10+G_g)}}\left(\frac{D}{D+2}\right)^{2T}=B_{liss}$$

P   = The couple's combined number of previous marriages

A_b = His age in years

A_g = Her age in years (biological, not cosmetic)

G_b = In millions, the number of hits when Googling his name

G_g = In millions, the numbers of hits when Googling her name

S   = Of her first five Google hits, the number showing her in clothing (or lack thereof) designed to elicit libidinous thoughts

D   = Number of months they knew each other before getting married (enter a fraction if necessary)

T   = Years of marriage. To find the likelihood of their marriage surviving 1 year, enter 1; for the likelihood of it lasting 5 years, enter 5, etc.

B_{lis} is the percentage chance that this couple's marriage will last for the number of years you chose.

While Sundem's celebrity marriage formula was less than serious, it nonetheless proved immensely popular, as well as remarkably prescient. Using it to chart a string of recent nuptials involving the rich and famous, some conclusions were to be expected. For example, Prince William and Kate, the Duke and Duchess of Cambridge, turned out to have a far better chance of being a long-term item than, say, Khloe Kardashian and Lamar Odom—the latter of whom announced their divorce in December 2013. But the formula also correctly surmised that, while Katie Holmes and Tom Cruise's marriage would last five years, there was little to no possibility it would reach fifteen. (Holmes and Cruise announced their divorce in 2012, following five and a half years as husband and wife.) The same fate was predicted for Will Smith and Jada Pinkett, who have passed their five-year anniversary, but who are regular tabloid fodder predicting impending doom. Meanwhile, the formula suggested that if "Brangelina"—Angelina Jolie and Brad Pitt—had married immediately after Pitt divorced previous wife Jennifer Aniston, there would have been a 1 percent chance of their marriage lasting 15 years. Today, their chances of doing so stand at more than 50 percent.

By Sundem's own admission, he is not a serious mathematician. The success of his pop formulas speaks less to the rigor of his empirical approach than to the general public's overwhelming eagerness to find answers to replace those that in a less technological society might have been chalked up to fate. If there is one absolute truth that comes out of his celebrity marriage equation, it is not that millions of Google hits doom a relationship from the very start, but that as people we are spectacularly bad at predicting who they are going to be well matched with when it comes to marriage.

After all, if the wealthiest, most successful and best-looking 1 percent of the population (i.e., celebrities) can't guarantee that they will be happy with their chosen marriage partners, what hope do the remaining 99 percent of us have? About half of first marriages fail in the United States, as do two-thirds of second marriages and three-quarters of third marriages. That we are extraordinarily bad at choosing our romantic partners in marriage shouldn't come as a great surprise. If there is one thing that we are told over and over again in movies, songs and novels, it is that love is fundamentally unpredictable by its very nature. Take Wagner's 1870 opera *Die Walküre*, for instance, in which the lovers turn out to be long-lost brother and sister, although this fact alone is not enough to stop them falling in love. Or consider the immortal words of pop singer Joe Jackson: "Is She Really Going Out with Him?"

In many cases, the unpredictability of love has rendered even the most scientific of minds incapable of explanation. "The heart has its reasons, which reason knows nothing of," Blaise Pascal, the French mathematician and inventor of the mechanical calculator, famously proclaimed. At the start of his 1822 book *On Love*, the French writer Stendhal lays out his desire to describe "simply, rationally, mathematically . . . the different emotions which . . . taken together are called the passion of Love."[3] The results, at least according to fellow wordsmith Henry James, were "unreadable." The problem, at least for those in the natural sciences, may be that there is not necessarily a natural law as relates to love. It might be eternal, but is it also external? Maybe not. "Love is intrinsically ametric," observed the British psychotherapist David Brazier, speaking of love's refusal to succumb to quantification.[4] Compounding the problem yet further is the suggestion that the merest attempt to analyze pleasure or beauty all but destroys it—not dissimilar to trying to get a closer look at an exotic bird, only to have it fly away when it senses our designs.[5]

This may be a defeatist approach, however, which is why two centuries after Stendhal's failed attempt to describe love, technologists continue to try to discover love's code—perhaps taking at face value Friedrich Nietzsche's assertion that "there is always some madness in love, but there is always some reason in madness."[6] And when it comes to this nobly scientific pursuit (worth upward of $4 billion globally in online dating fees alone[7]) there are few better people to examine than Neil Clark Warren and his eHarmony empire.

Seeking Harmony

Judged on first impressions, Neil Clark Warren is one of the more unlikely entrepreneurs of the Internet age. Born September 18, 1934, Warren was 65 years old when he launched his Internet dating company, eHarmony, in 2000. In the world of high tech, where acne is good, wrinkles are bad, and programmers throw in the towel at 30 because "coding is a young man's game," 65 might as well have been *165*. Warren wasn't—and isn't—a computer whiz. According to him, he didn't even know how to send and receive e-mail until his company was up and running. "I'm confessing up front that I don't know much of anything about algorithms," he says soon after he and I begin our conversation. "I think you're going to be terribly unimpressed."

But if Warren is too old to have grown up as part of the computer revolution, he was just the right age to have lived through another major change in American life. On October 28, 1935, when Warren was barely one year old, the *New York Times* ran a story in which it reported that the country's divorce rate had increased by more than 2,000 percent. Worse, the statistics didn't tell the whole story: according to the article, half of all American couples were supposedly unhappy; living, as one expert phrased it, "not really married but simply undivorced . . . in a sort of purgatory."[8]

As is often the case with moments of profound social

change, the tipping point that had led to this crisis of marriage wasn't limited to one factor but was rather the result of a number of different factors converging simultaneously. Love was being increasingly emphasized over duty in marriage—with the implication that matrimony should be the fount of all human happiness. For the first time, divorce was also a real option to those outside of elite society. America's cities were growing bigger by the day, as people moved away from their small, tight-knit communities and experienced a smorgasbord of new, previously unimaginable possibilities. The country was similarly more secular, as religion's influence receded and The Formula's advanced.

Right on cue came the new field of sociologists: invoking the metaphor of marriage as a machine, to be kept well oiled and in good working order at all times. As Paul Popenoe, one of the founding practitioners of American marriage counselling, observed:

> There is nothing mysterious about [marital relations], any more than there is about the overhauling of an automobile that is not working properly. The mechanic investigates one possibility at [a] time: he checks the ignition, the carburettor, the transmission, the valves, and so on; finds where the trouble lies; and removes the cause if possible. We do the same with a marriage.[9]

Warren's own parents fell into the category of troubled marriages. Although they stayed married for 70 years, Warren never felt that they were well matched with one another. "My dad was just so stinking bright, and my mom was so sweet, but she was two standard deviations below him in intelli-gence," he would recall. Warren's father—a business-minded man who owned a Chevrolet dealership, a John Deere outfit and a food shop—was interested in world politics, constantly questioning issues like the ongoing conflict between the Jews and Arabs. His mother, Warren has said, "didn't know there *was* a Middle East." When his father ran for office in Polk County, his mother voted for the other candidate.[10]

"I originally planned to go into the ministry, but discovered that I'm more of an entrepreneur than I am a minister," War-ren says. After earning a PhD in clinical psychology from the University of Chicago in 1967, he started out working as a marriage counselor. In 1992, he found fame by publishing a book called *Find the Love of Your Life*. It sold around a million copies, but Warren found himself feeling despondent. Even premarital counseling, he had concluded, was leaving things too late for many couples. "People have a tendency to form bonds very early in their relationship, and no matter what they find out at that point about whether they should get married or not, they will very seldom break up," he says. "I came to realize that the only way to match people up with one another is to do it before they have met and got involved."

Unlike a lot of psychologists, Warren discovered early in his career that he was interested in the quantitative parts of the field that scared off most people. He decided to put this to work by investigating the quality of marital relations. After *Find the Love of Your Life*, Warren started work on a large-scale research project, in which he conducted in-depth inter-views with 800 couples. When this was done, he compared the results of the 200 most satisfied and least satisfied couples and used this to derive a robust set of psychometric factors he found remarkably predictive of compatibility in marriage. "The results overlapped so much with what I had found from 40 years of therapy that I became more and more confident that I knew what I was talking about," he says. To help make sense of the data, he called in the services of a young statistician named Steve Carter—who is today eHarmony's Vice President of Matching. On one occasion early in the two men's working relationship, Warren threw out a curveball.

"How about building a website?" he said.

"To test for marital quality, you mean?" Carter answered.

"Not exactly," Warren countered. "I was thinking more along the lines of a matchmaking site. We could use our compatibility models as a start and go from there."

Despite some initial misgivings, Carter began working on the site: diligently reinterpreting the data Warren had gathered as a 436-question personality profiling test. Matching was done according to a checklist called the "29 Dimensions of Compatibility." These dimensions were composed of two main variables described as Core Traits ("defining aspects of who you are that remain largely unchanged") and Vital Attributes ("based on learning and experience and are more likely to change based on life events and decisions you make as an adult"). Although the algorithms involved remain largely black-boxed to avoid scrutiny, a number of details have come to light. Women, for instance, are never matched with men shorter than themselves, which means that men are similarly never paired with taller women. There are age parameters, too, which vary with gender, so that a 30-year-old woman will be matched with men between the ages of 27 and 40, while a 30-year-old man will be matched with women between 23 and 33.

In all, eHarmony's arrival represented more than just another addition to an already crowded field of Internet dating websites—but a qualitative change in the way that Internet dating was carried out. "Neil was adamant that this should be based on science," Carter says. Before eHarmony, the majority of dating websites took the form of searchable personal ads, of the kind that have been appearing in print since the 17th century.[11] After eHarmony, the search engine model was replaced with a recommender system praised in press materials for its "scientific precision." Instead of allowing users to scan through page after page of profiles, eHarmony simply required them to answer a series of questions—and then picked out the right option on their behalf.

The website opened its virtual doors for the first time on August 22, 2000. There were a few initial teething problems. "Some people were critical of the matches they were getting," Warren admits. "One woman who worked for a public relations firm was annoyed because we had matched her with two truck drivers, one after the other. My point was, 'You know, a truck driver can be very smart,' but in her mind there was such a status difference that she just thought it was absurd. She felt we weren't doing our job at all."

The point where, in Steve Carter's words, "shit took off" was when Jay Leno put on a wig and started spoofing Neil Clark Warren on his television talk show. "He was fascinated that this old guy, with gray hair, was talking about matching up people for marriage," Warren says. Although he was embarrassed, this was the moment he knew the website had struck a chord. By the end of 2001, eHarmony had more than 100,000 members registered. By August the following year, it had 415,000 members. By November 2003, the number had jumped to 2.25 million; followed by almost 6 million by December 2004; 8 million by October 2006; and 14 million by March 2007. Today, eHarmony credits its algorithms with generating 600,000 marriages in the United States—and a growing number overseas.[12] As it turned out, there was a demand for scientific approaches to matchmaking, after all. "From my point of view, I can't imagine anything I could be doing using statistics that would be more impactful on society than what I'm doing," Carter says.

Categorize Your Desire

eHarmony doesn't present itself simply as another coffee shop, nightclub, workplace or anywhere else that we might meet potential spouses. In its own words, it is designed to "deliver more than just dates" by promising connections to "singles who have been prescreened on . . . scientific predictors of relationship success." To put it another way, the person who uses eHarmony's algorithm to find dates isn't just being offered *more* dates, but *better* ones.

Certainly, Internet dating offers quantitatively better odds of finding a partner. Upward of 30 percent of the 7 billion people currently alive on Earth have access to the Internet, with this figure rising dramatically to around 80 percent in the United Kingdom and North America, where usage is among the highest. Where previously the romantic "field of eligibles" available to us (as sociologist Alan Kerckhoff once phrased it) was limited to those people we were likely to come into contact with as part of our everyday social network, the Internet now provides access to an unprecedented quantity of people, numbering around 2 billion in total.[13] Of these, a growing percentage have shown interest in using the Internet to find love. In a typical month, almost 25 million unique users from around the world access some form of online dating site.[14]

Does it offer a qualitatively better experience, however? This is, after all, the mission statement of websites like eHarmony—which jealously guard not only their user base of customers, but more importantly the proprietary algorithms used to match them. Generally speaking, sites that offer matching services tend to rely on similarity algorithms (with a certain amount of complementarity matching added in). Partners are matched on numerous personality traits and values, reflecting not only what the creators of a particular algorithm deem important, but also what is prized by its customers. This, in turn, has allowed a broad range of matching sites to pop up around eHarmony, each one catering to a different clientele who stress the importance of different aspects of the romantic experience.

For those interested in playing the numbers game, there is OkCupid, whose tagline borrows from eHarmony's in-your-face scientism by promising that "we use math to get you dates."[15] For the ultra-scientific (or perhaps the latent Eugenist) there is GenePartner.com, a website that claims to have "developed a formula to match men and women for a romantic relationship based on their genes."[16] For the low, low price of just $249, would-be daters can order a kit with which to swab the inside of their mouths. This saliva sample is then mailed back to the lab at GenePartner, where it is analyzed, and the results fed back to the user. For an additional fee, GenePartner can even match genetically similar users with their supposed soul mates. "With genetically highly compatible people we feel that rare sensation of perfect chemistry," the company's press materials state. "This is the body's receptive and welcoming response when immune systems harmonize and fit well together. Genetic compatibility results in an increased likelihood of form-ing an enduring and successful relationship, a more satisfying sex life, [and] higher fertility rates."

For the narcissistic, meanwhile, there is FindYourFaceMate .com, which hails itself as a "revolutionary new online dating site that employs sophisticated facial recognition software and a proprietary algorithm to identify partners more likely to ignite real passion and compatibility." Based on the theory that we are innately drawn to those people with features that resemble our own, subscribers to FindYourFaceMate upload profile pictures, which are then analyzed on nine different facial parameters (eyes, ears, nose, chin and various parts of the mouth) in order to find suitable "face mates."[17]

Then there is BeautifulPeople.com—described by its founder as a "gated community for the aesthetically blessed"—which marries skin-deep shallowness with a striking free-market ideol-ogy, summed up by its promise that "beauty lies in the eyes of the voter."[18] For the plus-sized dater, there is LargeAndLovely .com ("Where Singles of Size & Their Admirers Meet") and, on the other end of the spectrum, FitnessSingles.com ("Where Relationships Workout"). There are sites like SeaCaptainDate

.com ("Find Your First Mate") and TrekPassions.com ("Love Long & Prosper"); those that match based upon your particular strain of vegetarianism (VeggieDate.org); and those that pair people on a shared taste in books (ALikeWise.com) or a fondness for those who work in uniform (UniformDating.com).

The idea, essentially, is to drill down until we discover the particular weighted node that best captures our fancy. For some, their vegetarianism might be a defining characteristic and, therefore, a "must have" demand. For others it is nonessential, or even incidental. Online, not only is everyone a formula, as argued in the previous chapter, there is also a formula for everyone.

Love in the Time of Algorithms

There is a scene in the 2009 comedy film *Up in the Air* in which Natalie, an upwardly mobile businesswoman just out of college, describes the qualities she is looking for in a partner. She calls this her "type." "You know, white collar," she says, listing her ideal mate's attributions. "College grad. Loves dogs. Likes funny movies. Six foot one. Brown hair. Kind eyes. Works in finance but is outdoorsy, you know, on the weekends. I always imagined he'd have a single-syllable name like Matt or John or Dave. In a perfect world, he drives a 4Runner and the only thing he loves more than me is his golden Lab. Oh . . . and a nice smile."

We laugh at the scene because of the speed and exactness with which Natalie is able to recite all of these details. But the joke also rings true because all of us have met someone like Natalie. We may even *be* someone like her.

In one sense, the idea that selecting the perfect lover is the result of finding the individual whose list of attributes best measures up against our own wish list seems like the most natural thing in the world. Whether we are looking for a spouse, a holiday or a new laptop, all of us make mental lists of the qualities we are searching for and then match up whichever potential offerings we come across with our checklist of minimum demands. If a potential relationship is deemed not attractive enough for us on some level, a holiday is too expensive, or a laptop won't carry out the tasks we are buying it for, we dismiss it and move on to the next option. However, is this really the right way to think about love? In his book *How the Mind Works*, the experimental psychologist and author Steven Pinker poses a question very similar to the one asked in the Stable Marriage algorithm described at the start of this chapter: namely, how can a person be sure in a relationship that their partner will not leave them the moment that it is rational to do so? Pinker gives the potentially problematic example of a more physically attractive "10-out-of-10" neighbor moving in next door to us. The answer economists David Gale and Lloyd Shapley would give us is that we are in the clear just so long as this neighbor is already paired with someone more preferable to themselves than our spouse. The answer Pinker instead presents us with is the more humanistic suggestion that we ought not to accept a partner who wants us for any rational reason to begin with—but rather who is committed to staying because of who we are.

Murmuring that your lover's looks, earning power, and IQ meet your minimal standards would probably kill the romantic mood, even though the statement is statistically true. The way to a person's heart is to declare the opposite— that you're in love because you can't help it.[19]

This antirational view of love is one that may be breaking down in the age of The Formula. In *Love in the Time of Algorithms*, author Dan Slater relates the story of "Jacob," a thirty-something online dater who meets a 22-year-old secretary named Rachel on the Internet. After dating for a few months and moving in together, the couple split up after coming to the conclusion that they want different things from life. So far, so normal. What is different, however, are the comments made by Jacob in the wake of the relationship's failure. "I'm about 95 percent certain that if I'd met Rachel offline, and if I'd never done online dating, I would've married her," he tells Slater. "At that point in my life I would've overlooked everything else and done whatever it took to make things work. Did online dating change my perception of permanence? No doubt. When I sensed the breakup coming, I was okay with it. It didn't seem like there was going to be much of a mourning period, where you stare at your wall thinking you're destined to be alone and all that. I was eager to see what else was out there." In other words, there are plenty more fish in the sea—or, in the words of dating website PlentyofFish, at least "145 million [unique] monthly visitors."

Are You Sure You Want to Delete This Relationship?

Jacob's words offer a perfect summation of what the Polish sociologist Zygmunt Bauman refers to as "virtual relationships" in his book *Liquid Love: On the Frailty of Human Bonds*. Categorizing a more scientific approach to love alongside quick fixes, foolproof recipes, all-risk insurance and money-back guarantees, Bauman describes how the ultimate promise of virtual relationships is to "take the waiting out of wanting, [the] sweat out of effort and [the] effort out of results."

> Unlike "real relationships," "virtual relationships" are easy to enter and to exit. They look smart and clean, feel easy to use and user-friendly, when compared with the heavy, slow-moving, inert messy "real stuff." A twenty-eight-year-old man from Bath, interviewed in connection with the rapidly growing popularity of computer dating at the expense of singles bars and lonely heart columns, pointed to one decisive advantage of electronic relation: "You can always press 'delete.'"[20]

It is explanations such as this suggestion of "liquid love" that might mean—in the words of FreeDating website founder Dan Winchester—that the future will be made up of "better relationships but more divorce." Although this seems a paradoxical statement, it is something that could be the end result of better and better algorithms, Winchester suggests. "I often wonder," he says, "whether matching you up with great people is getting so efficient, and the process so enjoyable, that marriage will [eventually] become obsolete."

What Winchester is expressing is not unique, although it might well be a new phenomenon. In his 2004 book, *The Paradox of Choice: Why More Is Less*, the American psychologist Barry Schwartz argues that the overwhelming amount of available choice in everything from shopping to, yes, dating has become for many people a source of anxiety in itself.[21] In terms of relationships, this "paradox of choice" is dealt with by subjecting individual lovers to segmentation: an industrial term that denotes how efficiency can be gained by dividing up and isolating the means of production. In an age of mass customization, relationships become just any commodity to be shaped according to fads, changing desires and flux-like whims.

Such a postindustrial approach to dating runs counter to what we have been culturally conditioned to believe. The Lover is supposed to be unique; not just a combination of answers to set questions, each one to be answered correctly or incorrectly. Anyone who has ever scanned through page after page of Internet dating "matches," however, will soon find themselves, to quote the French novelist Marcel Proust, "unable to isolate and identify . . . the successive phrases, no sooner distinguished than forgotten."[22] Profiles become a seemingly never-ending repetition of terms like "cute," "fun-loving," "outgoing," "romantic" and "adventurous": each striving to break the formulaic mold of uniformity, but all ending up drawing from the same cultural scripts of desirable characteristics nonetheless. A similarly autistic approach to human relations can come across in the profile-building questions asked by dating websites. "Does money turn you on less or more than thunderstorms?" "What about body piercings versus erotica?" "Would you rather your potential spouse has power or long hair?" Since everything is given equal weighting on an interface level, and so many of the algorithm's inner workings are obscured, there is no way of knowing just what affects our final score.

This issue is one that was touched upon in a 2013 article for the *Guardian*'s "Datablog" (whose tagline reminds us that "Facts are sacred"). Recounting her experiences of algorithmic matchmaking, journalist Amy Webb writes, "When I first started online dating, I was faced with an endless stream of questions. In response, I was blunt, honest, and direct. Then my patience started to wear thin, so I clicked on what I thought sounded good." Before long, Webb began to second-guess the answers that she was being asked to enter. Certainly, she liked strong men who work with their hands, but was this a veiled attempt to ask whether she would date a lumberjack? After all, "they're strong and work with their hands," she says. "But I don't want to marry a lumberjack. I don't even like trees that much."[23]

Perhaps even worse is the existential crisis that results from a seemingly objective algorithm determining that it has scoured the Internet and that there is no one out there for you. "About 16 percent of the people that apply to us for membership we don't allow to participate on our site," Neil Clark Warren says. "We have seven different reasons for excluding people. If they are depressed—because depression is highly correlated with other pathologies—we don't let people participate. If they've been married more than three times we don't let them participate, which eliminates 15.5 percent of the marriages in America, which involve at least one person that's been married three times previously. In addition, we have something called 'Obstreperousness,' who are people that you just can't satisfy. You give them one person and they'll criticize them for being too assertive. You give them another and they'll claim they're too shy. We have ways of measuring that, and we ask that those people don't continue on the site."

Taking the Chance Out of Chance

Gary Shteyngart's 2010 novel *Super Sad True Love Story* is a coming-of-age fable set in a dystopian New York City of the near future. Because of the world's total informational transparency, no scrap of personal information is kept secret. All a character has to do—as occurs during one scene in which the novel's bumbling protagonist, Lenny Abramov, visits a Staten Island nightclub with his friends—is to set the "community parameters" of their iPhone-like device to a particular physical space and hit a button. At this point, every aspect of a person's profile is revealed, including their "fuckability" and "personality" scores (both ranked on a scale of 800), along with their ranked "anal/oral/vaginal" preferences. There is even a recommender system incorporated, so that a user's history of romantic relationships can be scrutinized for insights in much the same way that a person's previous orders on Amazon might dictate what they will be interested in next. As one of Abramov's friends notes, "This girl [has] a long multimedia thing on how her father abused her . . . Like, you've dated a lot of abused girls, so it knows you're into that shit."[24]

The world presented by *Super Sad True Love Story* is, in many ways, closer than you might think. Several years ago, the Human Dynamics group at MIT created a mobile system designed to sound an alarm if it was within ten yards of an attractive potential date. According to two of the researchers, Nathan Eagle and Alex Pentland, the system was "developed to enable serendipitous interactions between people" and was thus given the name Serendipity. As per the promotional literature supplied by the team:

In a crowded room you don't even have to bother working out who takes your fancy. The phone does all that. If it spots another phone with a good match—male or female—the two handsets beep and exchange information using Bluetooth radio technology. The rest is up to you.[25]

Apps like Serendipity are part of a new trend in technology called social discovery, which has grown out of social networking. Where social networking is about connecting with people already on your social graph, social discovery is all about meeting new people. There are few better examples in this book of The Formula in action than MIT's Serendipity system. Here is a problem ("chance") and a task ("making it more efficient"). Executed correctly, the computer might provide an answer to the question asked by Humphrey Bogart's character in *Casablanca*. "Of all the gin joints in all the towns in all the world, why did such-and-such a person walk in to yours?" Because the device in their pocket beeped.

Naturally, this raises as many questions as it answers. Serendipity is the occurrence of events by chance in such a way that winds up being beneficial to those involved. Equal weighting is placed on both the "random" and "beneficial" nodes, the absence of either one meaning that no matter what occurs, we may say with some certainty that it is no longer serendipitous. The distinction becomes more pointed when love is involved. After all, for many people the essence of love (if it can be described as such) is the forging of a certain universal value out of pure randomness: the idea that an apparently meaningless, chance encounter brings with it the ultimate meaning.

One of the best literary analyses of this phenomenon comes

in Alain de Botton's debut novel, *On Love*, in which the narrator becomes smitten by a woman he meets on a Paris-to-London flight. In an effort to inject a degree of rationality into the irrational, de Botton's protagonist tries to calculate the odds of such a meeting taking place, eventually coming up with an answer of 1 in 5840.82. "And yet," he writes, "it happened."

> The calculation, far from convincing us of the rational arguments, only backed up the mythical interpretation of our fall into love. If the chances behind an event are enormously remote, yet the event occurs nevertheless, may one not be forgiven for invoking a fatalistic explanation? . . . [A] probability of one in 5840.82 [makes it seem] impossible, from within love at least, that [our meeting] could have been anything but fate.[26]

Of course, as the late Steve Jobs might have said about fate: "There's an app for that." Serendipity's creators proudly state, "Technology is changing the way we date. For the shy and single, it has been the biggest aid to romance since the creation of the red rose."

Wear Your Heart on Your Sleeve

MIT's Serendipity project is not the first investigation of its kind. In the late 1990s, a proximity matchmaking device called Lovegety briefly became all the rage in Japan, selling more than 1.3 million units at an approximate price of $21. The aim of Lovegety was to allow users to find potential dates in their vicinity. Users were asked to input responses to several questions, which in turn became their personal list of preferences. When a mutual match was discovered within range, the device alerted both parties to one another's presence. Compared to eHarmony's barrage of questions and 29-point compatibility scale, Lovegety's preference options were admittedly sparse. Even with this being the case, however (and despite the preferences on offer being extremely superficial, such as a shared desire to partake in karaoke), a surprising number of users still described the experience as qualitatively different from that of being approached by a stranger. "When you're picked up out of the blue, there is always an element of suspicion," said one female customer. "But when you're brought together through the Lovegety, you're more at ease because you already have something in common. You already have something to talk about."[27] In other words, the technology was more than just an invisible mediator between two parties. Like the shared ownership of a Porsche, being part of the Lovegety club meant there was an automatic commonality between both parties. To paraphrase Marshall McLuhan, the medium really *was* the message.

Such devices don't have to remain the sole province of bored teenagers in Tokyo's Harajuku district, of course. In the Republic of Iceland, the Islendiga-App ("App of Icelanders") applies similar technology to the problem of solving the issue of accidental incest in a country of 320,000, where almost everyone is distantly related to one another. By accessing an online database of residents and their family trees stretching back 1,200 years, and then using an algorithm to determine the shortest path between two points, the app is

able to inform users how closely related they are to the person
they might be considering sleeping with. The app is activated
by "bumping" phones, which in turn triggers an "incest pre-
vention alarm" in situations where the number of genealogical
steps is sufficiently small as to cause potential distress at future
family gatherings. In the words of the app's admittedly catchy
slogan: "Bump the app before you bump in bed."[28]

Today, a large number of proximity-based social-discovery
apps—from Grindr to Crowded Room—vie for prominence
in the marketplace. Among the more interesting of those I
came across during my research was Anomo, an app that lays
out its ambitions to "democratize the way we socialize" in a
way that combines both the actual and virtual worlds.
Cofounded by a former runner-up on the U.S. version of *The
Apprentice*, Anomo asks its users to create video-game-style
avatars for themselves, which then become their virtual prox-
ies as they negotiate their way through the actual world. "A
lot of existing social discovery apps are about real profiles,"
says cofounder Benjamin Liu. "You put your real picture and
name on there. The problem with that is that people's first
reaction to it is often, 'Wow, that's really creepy.'" Anomo
users do enter their real information to the app, but this is
hidden until users choose to reveal it in a series of quid pro
quo, "you show me yours, I'll show you mine" exchanges.
Prior to that, users have the option of either chatting to one
another using their avatars (which consist of a short description, and one of around 100 different cartoon representations), or else playing a series of "icebreaker" games to
determine how much they have in common with a person
without having to endure a potentially awkward chat first. By

comparing responses to questions such as "Do you prefer beer or wine?" or "Is a bicycle built for two cheesy, cool, or dangerous?," Anomo's app provides a "compatibility rating" ranking your answers next to those given by the other person. "Interacting anonymously changes the playing field," states Anomo's promotional literature. "Suddenly first impressions are not based on photos, but via a genuine connection."

Tapping the Scene

For the most part, social discovery works along the same lines as Instant Messenger. Rather than focusing on finding "the one" users are instead dealing with a pool of available participants that exists in a constant state of flux. It doesn't matter if a particular message is not returned, or the person we originally wished to "chat" with is no longer present. A person simply moves to the next user in line and begins the process over again, since there are always enough people online (at least in the case of those apps that find success) to counteract the loneliness of any given moment.

While most of these apps require that people consciously engage with them, this rule is by no means an absolute. In his most recent book, *Who Owns the Future?*, computer scientist and virtual reality pioneer Jaron Lanier recalls a panel he served on at UC Berkeley, judging start-up proposals submitted by graduate engineering students enrolled in an entrepreneurial program. A group of three students presented a concept for quantifying nights out so as to ensure maximum romantic success for those involved. "Suppose you're darting around San Francisco bars and hot spots on a Saturday night,"

Lanier remembers the group pitching. "You land in a bar and there are a bounteous number of seemingly accessible, lovely, and unattached young women hanging out looking for attention . . . Well, you whip your mobile out and alert the network."[29] Such an idea would, of course, never work, Lanier observes, since the data would invariably be inaccurate and the scheme would wind up running on hope.

His words of warning have not been enough to derail the technology's core concept, however. SceneTap—previously known as BarTabbers—started life in Chicago, although it has since expanded to cover San Francisco, Austin, Columbus, New York, Boston and Miami. With cameras installed in more than 400 drinking establishments, SceneTap uses facial-recognition technology and people-counting algorithms to help bar-hoppers decide which locations to hit up on a particular night out. Currently, the tool can provide real-time information on crowd sizes, gender ratios and the average age of patrons in any given location—although that is not everything that's planned. In 2012, the start-up filed a patent that, in the words of Forbes.com, "[crosses] the creepy line . . . and then keeps running and spikes the creepy football in the creepy end zone."[30] In a nutshell, the patent is designed to allow SceneTap to collect much more detailed data, including bar-goers' race, height, weight, attractiveness, hair color, clothing type and the presence of identifying features like facial hair or glasses. In other words, one might access the app and be presented with information along the lines of: "The Raven is 73% full. Its crowd is made up of 33% natural blondes, 57% brunettes, 3% redheads, 5% bottle blondes, and 2% other. The men on average are 5.8 feet tall, and 70% are dressed in business casual.

Women on average weigh 154.7 pounds, and 24% wear short skirts. 73% of patrons are white, 21% Asian and 6% black. Attractiveness average for the location is 7 out of 10."

The patent also allows for microphones to be placed in cameras in order to pick up on what customers are saying, as well as for the facial recognition technology to identify people and link them with their social networking profiles to determine "relationship status, intelligence, education and income."

No doubt STD tracking could be added at a later date.

Your Sex Life with Models

"I've been in a relationship with just one woman my entire adult life, ever since I was in high school," says Kevin Conboy, a soft-spoken computer programmer with a mop of brown, curly hair and a Brian Blessed beard. About ten years ago, when Conboy was in his mid-twenties, working as a user-interface engineer, he tried to work out how many times he and his wife had had sex in their time together. The idea grew in scale and, before long, Conboy was spending his evenings working on an application to keep tabs on his sex life: methodically cataloguing and modeling everything from the duration and frequency of sex to its quality and average time frame.

"I strive for honesty in my life as much as possible, but I thought that this was one thing that was better to build in secret and then to show my wife once it was completed to ask for permission," he says. "At first, she thought it was kind of weird, but she also understood that to me code is a form of self-expression. She asked me for the link to view it, and she would check up every now and then to see how our sex life was doing. It got to the point where she would tell her friends about what I had created, and it sort of became this talking point."

It was a friend of Conboy's wife who eventually convinced him to open up the app for public use. Giving it the name Bedpost, and the tagline "Ever wonder how often you get busy?," the app advises users to "simply log in after every time [they] have sex and fill out a few simple fields. Pretty soon, you'll have a rolling history of your sex life on which to reflect."[31]

This mass of data can be visualized in a variety of forms—including pie charts and scatter plots—with heat maps showing different intensities of color based on the quantity and quality of sex a particular user is having. "The amount of data you can attach to sexual activity is uncapped," Conboy says. "For instance, if you're on your phone, there is no reason you can't record the GPS data. You might be having sex all over the world and it could be fun to look back at all that information."

The goal of his work, Conboy says, "is to get you to think about your sex life in a way that you hadn't previously." Its success hints at something interesting: that the act of measuring and quantifying sex in the form of zeroes and ones can in itself become an erotic act. After all, if text-based cybersex based entirely around semiotic interaction can be arousing then why can't the code that underpins it?

When I spoke with Conboy for the first time, he was busy rethinking the Bedpost user interface, which had remained largely unchanged for several years. A number of the site's best suggestions, he said, had come from the app's users themselves. The popular feature of menstruation tracking, for example, was something he says he never would have come up with by himself. In particular, he was fretting over whether or not to add an orgasm counter to the list of available features. There was also the question of adding greater social-networking capabilities ("This might freak some people out, but I like the idea of being able to pull people in via your Facebook and Twitter accounts"), along with the ability for users' partners to log in and add their own "tags"—thereby creating a type of wiki criticism of "sex-as-performance," which could take place long after the act had taken place.

The ability to share the experience of sex is one that is particularly fascinating. What if we were able to get closer to the other's sexual experience not simply by asking, "How was it for you?" but by actually delving into the data for evidence of the other person's satisfaction? This wouldn't necessarily have to rely on anything as subjective as user tags. A number of pieces of wearable tech have already shown themselves to be uniquely valuable when it comes to quantifying the sex experience. By recording heart rate, perspiration output and motion patterns, the BodyMedia FIT armband can, for instance, recognize whether wearers are having sex at any particular time. More intrusively, it is even able to gauge whether wearers are having *good* sex or not, since analyzing spikes in the various metrics can reveal if a partner has faked an orgasm.[32]

Awkward conversations aside, this available data has the potential to raise a number of issues. For example, the spouse who suspects his or her significant other of cheating could access their data and question why 100 calories had been burned off between 0:13 and 2:00 A.M. on a particular evening, without their taking a single step, and with them falling asleep immediately afterward. Removing said tech (or else "forgetting" to wear it) might be grounds for suspicion in itself. Even arranging for illicit encounters to take place during daylight hours, when data anomalies would theoretically be more easily explained away, would be of little use since sexual activities look different in terms of fitness data than other energetic activities like weight-lifting, jogging, yoga, martial arts and cycling.

For Conboy, Bedpost helps to normalize him within the sexual spectrum. "The aggregate data gives me a sexual confidence," he says. "It's nice to know that you have had sex a certain number of times this month, or this year. It's a reminder that my sex life is healthy. The numbers help me to relax about whatever I'm obsessing over at the time. I don't think I could have an erotic sex life without the confidence the data gives me."

Of course, this raises yet more questions—not least how much data is enough?

No matter how scientific the intention, the moment we start measuring we also begin limiting, not just based on what the measurement tool is designed to capture, but on those metrics we consider worthy of measurement at the time. In the case of Bedpost, Conboy might add an orgasm counter, but how about respiratory rate? Or if he adds respiratory measurements, what about perspiration levels and heart rate? And so on it goes—the ideal and meaning of what we are measuring receding further and further from view, like advancing on an Impressionist painting and seeing what from a distance had been a convincing re-creation of a landscape dissipate into a sea of painted dots.

Night of the Loving Dead

If we accept the (highly questionable) view that true love, like opportunity, knocks only once, then how about the idea that love must automatically be separated by its deathly binary opposite? It has long been a technological pipe dream that man should be able to transport his "spirit" beyond the corporeal body and into the metallic lattices of a computer. In *The Enchanted Loom: Mind in the Universe*, NASA futurist Robert Jastrow waxes lyrical about his hopes for a future in which the human brain could be "ensconced in a computer . . . liberated from the weakness of the mortal flesh . . . in control of its own destiny."[33] As with natural birth, such a form of reproduction would ensure our immortality through the continuation of our perceived wisdom, good and happiness. To frame this as a question about love, what if apparent death didn't therefore part us from our loved ones but rather, as with a butterfly emerging from a chrysalis, simply meant a change in form factor?

It is into this space that applications such as IfIDie, DeadSocial and LivesOn enter the frame. All three apps exist in various magnitudes of complexity. DeadSocial and IfIDie function as what can essentially be described as netherworldly "out of office" services: allowing nothing more complicated than the writing and recording of timed Facebook messages, activated upon a particular user's death, and which can then be sent out to lovers, friends and family for years to come. In a video designed to promote the former service, we are told how the widow of a deceased DeadSocial user can even be "the recipient of an inappropriate joke from time to time."[34] IfIDie, meanwhile, recommends the recording of a webcam video in the event that one wishes to bid a fond farewell or (accompanied in literature by an image of a heavenly cloud in the shape of a raised middle finger) "settle . . . an old score" with a member of the living.[35]

More complex, both ethically and technologically, is LivesOn—which carries the distinctly *Twilight Zone* tagline, "When your heart stops beating, you'll keep tweeting."[36] LivesOn uses machine-learning processes to sift through your past social-network feeds, looking for the subjects, likes and news articles that interested you during life, so that similar subjects can be tweeted about on your behalf after death. Users are encouraged to help their LivesOn "train and grow" by following a person's existing Twitter feed and analyzing it, learning tastes, likes, dislikes and eventually even syntax. "The goal is to get it to almost become like a twin," says creator Dave Bedwood.

That isn't all. Several years ago, news broke that the U.S. Department of Defense was developing a project designed to create "a highly interactive PC or web-based application to allow family members to verbally interact with virtual renditions of deployed Service Members." Using high-resolution 3-D rendering, the high-level concept was to allow a child or spouse to engage in a simulated conversation with an absent parent or partner, and receive responses to stock phrases including "I love you" and "I miss you." "I'm not saying this kind of ghost is for everyone," commented one writer for *Slate*, with what is perhaps something of an understatement,

> but I dare you to tell a child who has lost her father in Iraq or Afghanistan that she can't keep a virtual rendition of him to help her go to sleep. And I dare you to stop the

millions of others who will want ghosts of their own when today's military project becomes, once again, tomorrow's mass market.[37]

While it would be the harsh writer who would willingly inflict pain on a person who has lost their loved one (although it seems odd to blame said writer for the absence of the parent/spouse to begin with), one cannot help but think that *Slate* has not properly thought through its position here. A computer program that is able to convincingly replicate the appearance, voice and sentence structure of an absent or deceased relative is neither quantitatively or qualitatively the same thing as having said person there—much as a dead body is not the same as a live one, even if it might look no different in a still photograph. To presume that the person we fall in love with, or even just the objects that hold personal value for us, are simply a composite of their individual properties is incorrect. Studies have shown that taking away the security blanket or teddy bear of a young child and substituting it with a duplicate will not see the replacement accepted in place of the original. We may well react in the same way if our Rolex watch was replaced with an unofficial replica, no matter how convincing a knockoff this might prove to be. Taken to extremes, it would be the remarkably unfeeling doctor who would reassure the parent of a dead child, or the widowed spouse of a dead husband or wife, by telling them that their loss needn't matter since we can clone the person in question— or better yet, that they have a twin who is still alive. In his writing, the French phenomenological philosopher Maurice Merleau-Ponty explored this idea, by distinguishing between what he views as the living and dead properties of an object. It is in The Formula's empirical desire to reduce perception and feeling to that which is observable, he argues, that all elements of "mystery" are lost.[38]

Desiring Machines

There is little disputing that Kari is an attractive girl. She has bright blue eyes, made all the more attention-grabbing by the smoky eye makeup she wears. Her hair is cut to shoulder length and is bottle blond in color, with just the slightest trace of what one presumes to be her natural brown creeping in at the roots. Her face suggests that she is in her early twenties, and as she looks toward the camera, her full lips part slightly as a coquettish smile plays over her features. It's only then that you notice for the first time that she is wearing cotton candy–colored lip gloss.

If Kari sounds a little, well, fake, that is entirely under-standable. She is not real, at least not in the sense that you or I are—existing purely in the form of computer code. Her name is an acronym, standing for Knowledge Acquiring and Response Intelligence, and she is what is known in the trade as a "chatterbot," an algorithm designed to simulate an intel-ligent conversation with an actual person. Kari (or, perhaps, K.A.R.I.) is the creation of Sergio Parada, a thirtysomething programmer living in West Palm Beach, Florida.

Born into a military family in San Salvador, El Salvador, Parada moved to the United States with his parents when he was ten in order to escape the Salvadoran civil war. After studying video games and programming at Chicago's Columbia College, Parada picked up a job working on the *Leisure Suit Larry* series of adult video games for the now-defunct developer Sierra Entertainment. It was while he was employed on the game tentatively titled *Lust in Space* (aka *Larry Explores Uranus*) that Parada came up with the idea of creating a piece of software centered entirely around one's interactions with a solo female character. "It was something brand new," he recalls, describing his aim as being "not just a girl simulation, but a relationship simulation." Unlike *Leisure Suit Larry*, which followed an affable loser as he attempts to bed a procession of ludicrously attractive women, in *Kari Virtual Girlfriend* there would be no set narrative. This necessarily meant that in the traditional game-playing sense there was also no way to win. Like a real relationship, the game continues indefinitely, with the reward being a relationship that grows and deepens.

In the same way that people anthropomorphize pets, attributing them with human characteristics to explain behavior, so too does Kari build on the willingness to protect, feed, shelter and guard machines that is second nature to a generation who grew up with Tamagotchis: the handheld digital "pets" that enjoyed a burst of popularity in the late 1990s. A number of psychology researchers (with two of the most prominent being Mihaly Csikszentmihalyi and Eugene Rochberg-Halton) have investigated the creation-of-meaning process that is called "psychic energy." The idea is that as a person invests more psychic energy in an object, they attach more meaning to it, thus making the object more important to them, which makes their attachment grow all the stronger. This effect is only strengthened by the fact that in Kari's case, she encourages the user to talk and proves an adept listener. This might suggest an additional reason for the magnetic pull that Kari wields over some users.

In a 1991 study by Arthur Aron, a psychology professor at Stony Brook University in New York, pairs of students who had never previously met were placed in a room together for 90 minutes, during which time they exchanged intimate pieces of personal information, such as their likely reaction to the loss of a parent or their most embarrassing personal moments. At the end of the time period, the two were asked to stare unspeaking into one another's eyes for a further two minutes. The students were asked to rate the closeness they felt at various intervals and, after just 45 minutes, these ratings outscored the closeness felt by 30 percent of other students, ranking the closest relationships they had experienced in their lives. Aron's conclusion was that disclosure is therefore both fast-acting and powerful as a device to increase personal attraction. (As if to bear this theory out, despite the students exiting the room from different doors after the experiment was over, so as to remove any sense of obligation to stay in touch, the first pair to take part were married six months later.)

One major difference in the relationship Kari has with her users, when compared to the vast majority of romantic relationships (aside from the obvious one), is that in order to avail oneself of her services, a person must first purchase her. The standard version of the program costs $24, while the "Kari Ultimate Package"—which comes with an Avatar Studio to "make your own girls"—will set users back $90.[39]

For those who do buy her, the relationship conforms to what French philosopher Michel Foucault refers to as "spirals of power and pleasure." The user's pleasure is linked to the power they hold over their virtual girlfriend. It is a technological reimagining of the *Pygmalion* fantasy, in which Kari takes on the role of Eliza Doolittle: the naive, young, working-class girl initiated into the sexual world by the older, wealthier, more experienced Professor Henry Higgins.[40] Of course, where Henry Higgins handcrafted his idealized companion by teaching her genteel manners and elocution, with Kari the user has the ability to fine-tune specific areas of her personality, such as her "love," "ego" and "libido" ratings, which are measured on an adjustable scale from 0 to 10. Find that Kari skips from topic to topic when she talks? Just increase her "stay on topic"-ness, or else change the number of seconds between unprovoked comments. Worry that she is starting to act too aloof? Lower her "independence" and, if that doesn't work, her "memory scope." This can have an unfortunately detrimental effect on Kari's AI, leading to one prickly reviewer complaining that "one minute she's spouting some dime store philosophy, the next she's asking to be 'f***ed so hard,'" and describing the result as "the video game equivalent of dating a drunken first-year philosophy student with really low self-esteem."[41]

Ultimately, it is Kari's malleability that forms the crux of her appeal. Unlike the other "flickering signifiers" we might fall in love with—be they film actor, pop star, or doe-eyed supermodel plastered across a billboard—Kari can adjust to appeal to each person individually; quite literally living up to the prostitute's classic sales pitch, "I can be whoever you want me to be."

"Kari fills that hole we all have inside ourselves for a connection with someone," Parada notes. "Some users write to me with word of the interesting projects they are conducting.

One wanted to get her pregnant and see the nine-month cycle of having a baby with his Kari. Another wanted to give her menstrual cycles. Another fed his entire journals to it and said it was better than therapy. Whatever you're up to, Kari is a big canvas to work with. It's a canvas of awareness and thought; an extension of the self. Whatever we choose to expose our Karis to is what they will become. It's like creating a being from the things we choose to teach it. The best part is that this is all integrated into an avatar that can talk back to you, be your best friend—and even love you."[42]

Love and Sex with Algorithms

The Kari-esque dream of an algorithmic lover has existed at least as long as there have been personal computers, from 1984's *Electric Dreams* (in which a San Francisco architect's newly installed computer system grows jealous of its owner's burgeoning relationship with an attractive neighbor) to Spike Jonze's 2013 film, *Her.*

In Japan, there exists a popular genre of video game known as the "dating sim." The typical dating sim (short for "simulation") allows the player to take control of an avatar, who is then confronted with a number of potential love interests whom they must choose between. As with role-playing games, dating sims rely heavily on statistics. A character's conversation with a would-be mate is measured according to their choice of appropriate lines of dialogue, with overall scores improving or worsening depending upon which line is chosen at a particular time. In one such scene, a female character is depicted eating an ice cream, a daub of it having smeared her cheek. Players are given the option of "Wipe it off for her" or "Pretend I didn't see it," while an additional piece of information warns, "Her affection level will change according to your choices."[43]

Perhaps the most notable exponent of this game-space view of relationships is artificial intelligence scholar and chess master David Levy. In Levy's most recent book, *Love and Sex with Robots*, he offers the prediction that not only will the concepts suggested in his title be possible in the future but that, by 2050, they will be routine. As Levy argues:

> One can reasonably expect that a robot will be better equipped than a human partner to satisfy the needs of its human, simply because a robot will be better at recognizing those needs, more knowledgeable about how to deal with them, and lacking any selfishness or inhibitions that might, in another human being, mitigate against a caring, loving approach to whatever gives rise to those needs.[44]

In a moment of particularly questionable human understanding, Levy theorizes that an algorithm could analyze the brain's "love measure" by way of an fMRI scanner and use this information to hone specific strategies to make a person fall in love. Like a game of chess, the path to true love would thus consist of a series of steps, each move either lowering or raising the love measure exhibited in the brain. In the manner of a smart central-heating thermostat, the aim would be to automatically adjust warmth or coolness so as to keep conditions at an optimal level. Suggestion that you watch a Tchaikovsky ballet together met with a frosty response? Try complimenting the other person's new haircut instead. An algorithm could even take note of the low-level features we might be struck by but unable to verbalize in partners, such as the way the lover flicks their hair or lights a cigarette, and incorporate this into its seductive repertoire.

Like Kari, Levy's proposed automated lover could even answer the old philosopher's question of how love demands that infinite desire be directed toward a finite object. Fancy a gawky, retiring lover one night and a "woman looking like Marilyn Monroe . . . with the brainpower and knowledge of a university professor and the conversational style of a party-loving teenager" the next? No problem. Rather than having to willfully idealize the human lover to make them unique, the ideal partner could be created, then constantly modified in the manner of a Facebook profile. Even the suggestion that we might be drawn to individuals with flaws doesn't appear to faze Levy. On the contrary, if a "perfect" relationship "requires some imperfections of each partner to create occasional surprises . . . it might . . . prove necessary to program robots with a varying level of imperfection in order to maximize their owner's relationship satisfaction," he writes. Later on, he gives an example of how this might practically function, as he imagines a lover's tiff in which the human party in the couple finally loses patience with their algorithm's operating efficiency and yells, "I wish you weren't always so goddamn calm." To resolve this issue and restore stability to its optimal level, Levy suggests that here it might be necessary for the algorithmic partner to simply "reprogram itself to be slightly less emotionally stable."

It's easy when you know how. Although one can't help but think you risk losing a certain spark in the process.

CHAPTER 3

Do Algorithms Dream of Electric Laws?

A decade ago, Walmart stumbled upon an oddball piece of information while using its data-mining algorithms to comb through the mountains of information generated by its 245 million weekly customers. What it discovered was that, alongside the expected emergency supplies of duct tape, beer and bottled water, no product saw more of an increase in demand during severe weather warnings than strawberry Pop-Tarts. To test this insight, when news broke about the impending Hurricane Frances in 2004, Walmart bosses ordered trucks stocked with the Kellogg's snack to be delivered to all its stores in the hurricane's path. When these sold out just as quickly, Walmart bosses knew that they had gained a valuable glimpse into both consumer habits and the power of The Formula.[1]

Walmart executives weren't alone in seeing the value of this discovery. At the time, psychologist Colleen McCue and Los Angeles police chief Charlie Beck were collaborating on a paper for the law-enforcement magazine *The Police Chief*. They too seized upon Walmart's revelation as a way of reimagining police work in a form that would be more predictive and less reactive. Entitled "Predictive Policing: What Can We Learn from Walmart and Amazon about Fighting Crime in a Recession?," their 2009 paper immediately captured the imagination of law-enforcement professionals around the country when it was published.[2] What McCue and Beck meant by "predictive policing" was that, thanks to advances in computing, crime data could now be gathered and analyzed in near-real time—and subsequently used to anticipate, prevent and respond more effectively to those crimes that would take place in the future. As per the slogan of Quantcast—the web-analytics company I described back in Chapter 1—it means that police could "Know Ahead. Act Before."™

Today, there is no man more associated with the field of predictive policing than Los Angeles Police Department's Captain Sean Malinowski. Despite his reputation within the force as a "computer guy," Malinowski's background is actually in marketing. Prior to his joining the LAPD in 1994, he worked as a marketing account executive, helping chewing gum and margarine companies roll out new products by figuring out how best to hit customers with targeted coupons and special offers. Through a stroke of good fortune Malinowski ended up working on a drunk-driving campaign with several officers from the City of Detroit Police Department. He found working with police to be a revelation. "Up until that point I was all gung-ho on the marketing thing," he says. "I hadn't thought about the police force before that. Part of it was that I had reached a time in my marketing career where I was thinking, 'Christ, is this it: my whole life's mission is going to be to sell edible fats and oils?' The cops I was working with had a real mission; they were trying to do something bigger."

Through some mutual friends, Malinowski was introduced to a former New York cop, who had recently moved to Chicago to work as an academic. After speaking with him, Malinowski quit the marketing business and went back to school. Several years later he graduated from the University of Illinois with a master of arts in criminal justice. Then it was on to the LAPD Training Division, where Malinowski wound up being elected class president. His first job with the police department proper was on Pacific Beach Detail, cycling endlessly up and down the Venice beachfront to interview street vendors and ensure public safety.

Malinowski's big break came when he was assigned to work for Chief William Bratton: first as his aide and later as his executive officer. Bratton had just moved to Los Angeles from New York City, where he'd established a formidable reputation as the cop who had halved the city's murder rate in the span of a few years. Bratton's methods, though effective, were undeniably offbeat. Prior to the New York Police Department, he had headed up the New York transit police, where he had transformed the New York subway system from a veritable hellhole into a clean, orderly microcosm of a law-abiding society by . . . *cracking down on fare dodging*. In other words, at a time when serious crimes on the subway were at an all-time high, Bratton focused his attention on making sure that people paid for their tickets. His reasons, as recalled in his 2009 memoir, were simple: fare evasion was a gateway drug to more serious crime. "Legitimate riders felt that they were entering a place of lawlessness and disorder," he noted. "They saw people going in for free and began to question the wisdom of abiding by the law . . . The system was veering toward anarchy."[3] By stopping and searching law violators for even the most minor of infractions, troublemakers decided that it was easier to pay their fares and leave their weapons (which were often uncovered during searches) at home. Crime fell exponentially.

Relocated to Los Angeles, Chief Bratton wanted to use some of that preemptive mojo on a grander scale. Working under him for five years, Malinowski witnessed firsthand how Bratton was able to take a department stymied by inertia and push through changes by sheer force of will. "When you're in a bureaucratic organization, you get so used to the barriers coming up that it can limit people's creativity," Malinowski says. "What Bratton instilled in me was not to be affected by all that. He taught me to think big and make things happen."

More than anything, Bratton was always on the lookout for the next "big idea" to revolutionize his work. In predictive analytics, he felt that he had found it. What Bratton had noticed was a correlation between crime rate and the speed at which that data could be analyzed. In 1990, crime data was collected and reviewed only on an annual basis. At the time, crime was on a steep rise in the majority of American cities. When 1995 rolled around, crime data could be looked at on a month-by-month basis. Crime rates during that same period slowed. With crime rates now viewable on a moment-to-moment basis, Bratton posited that this could lead to an actual drop in crime rates by predicting where crimes would next take place. In the same way that companies like Quantcast and Google are able to mine user data for insights, so too was the idea of predictive policing, that rather than simply identifying past crime patterns, analysts could focus on finding the next crime location within an exist-ing pattern. To put it in Amazon terms: *You stole a handbag; how about robbing a liquor store?*

Why Is Crime Like an Earthquake?

It is widely accepted that crimes don't occur randomly dis-persed across a particular area, but rather that they exist in small geographical clusters, known as "hotspots." In Seattle, for example, crime data gathered over a 14-year period shows that half of all crime can be isolated to just 4.5 percent of the city's streets. A similar observation holds true for Minneapolis, where 3.3 percent of all street addresses generate more than 50 percent of calls to police. Over a 28-year stretch, a marginal 8 percent of streets in Boston accounted for a whopping 66 per-cent of street robberies. Knowing about these hotspots, and the type of crime that is likely to take place at them, can be vital in helping direct police to specific parts of a city.[4]

Imagine, for example, that there are arrests for assault every Saturday night outside your local pub, the White Hart. If that proves to be the case, it wouldn't be too difficult to predict that future Saturday nights will see similar behavior at that location, and that stationing a police officer on the door at closing time could be enough to prevent future fights from breaking out.

It was this insight that prompted Chief Bratton to ask Sean Malinowski to help him.

On Bratton's advice, Malinowski started driving over to the University of California, Los Angeles every Friday afternoon to meet with members of the math and computer science depart-ments. The Los Angeles Police Department had agreed to hand over its impressive data set of crime statistics—which amounted to approximately 13 million crime incidents recorded over an 80-year period—for the largest study of its kind. Malinowski relished the experience of working with the UCLA researchers. As had happened a decade earlier when he first started working with police on his drunk-driving campaign, he found himself being drawn into the work being done by the computer scientists as they combed through the data looking for patterns and, hopefully, formulas.

"I loved those days," Malinowski recalls. Of particular interest to him was the work of George Mohler, a young mathematician and computer scientist in his mid-twenties, who was busy working on an algorithm designed to predict the aftereffects of earthquakes. Mohler's work was more relevant than it might initially sound. In the same way that earthquakes produce aftershocks, so too does crime. In the immediate aftermath of a house burglary or car theft, that particular location becomes between 4 and 12 times more likely to be the scene of a similar crime event. This is a type of contagious behavior that is known as the "near repeat" effect. "Often a burglar will return to the same house or a neighboring house a week later and commit another burglary," Mohler explains. Taking some of Mohler's conclusions about earthquakes— and with help from an anthropologist named Jeff Brantingham and a criminologist named George Tita—the team of UCLA researchers were able to create a crime prediction algorithm that divided the city into different "boxes" of around 0.15 kilometers squared, and then ranked each of these boxes in order of the likelihood of a crime taking place.

A three-month randomized study using the algorithm

began in November 2011. "Today . . . is historic," began Malinowski's address in that day's *Patrol Alert*. His division, known as Foothill, covered seven main beats: La Tuna Canyon, Lake View Terrace, Pacoima, Shadow Hills, Sun Valley, Sunland and Tujunga. When divided up, these amounted to 5,200 boxes in total. At the start of that day's roll call, Foothill patrol officers were handed individual mission maps, each with one or more boxes highlighted. These were the locations deemed as "high probability" and were accompanied by statistical predictions about the type of crime that was likely to occur there. "What we're asking you to do is to use your available time to get into those boxes and look around for people or conditions that indicate that a crime is likely to occur," Malinowski said, addressing his team. "Then take enforcement or preventative action to stop it."

The experiment ran until February the following year. In March, the results were evaluated, and a decision was made about whether or not to roll out the technology. Findings were impressive. During the trial, Foothill had seen a 36 percent drop in its crime rate. On days where the algorithm was dictating patrols, predictions about which crimes were likely to take place were twice as accurate as those made by a human analyst. "Part of the reason is that human brains are not suited to rank 20 hotspots across a city," George Mohler says. "Maybe they can give you the top one or two, but at seven or eight they're just giving you random guesses."

If there was a teething problem to all of this, it often came from Malinowski's own men. "You do run into people who say they don't need a computer to tell them where the crime is," he admits. "A lot of guys try to resist it. You show them the forecasts and they say, 'I could have told you that. I could have told you that the corner of Van Nuys and Glenoaks has always been a problem.' I say, 'That's always been our problem, huh? How long have you been working here?' They go, 'Ten years I've been working this spot.' And I say, 'Then why the hell is it still a problem if you've known about it for ten years? Get out there and fix it.'"

Following the Foothill study, algorithmic policing was made available to all Los Angeles precincts. Similar algorithms have now been adopted by other police departments around the United States. Malinowski says that he still feels responsible for his officers but is getting used to his less hands-on role in their deployment. "You have to give up a bit of control to let the algorithm do its work," he notes. Chief Bratton, meanwhile, retired from the Los Angeles Police Department. Following the 2011 England riots, he was approached by David Cameron about coming to the UK as Commissioner of London's Metropolitan Police Service. The offer was ultimately vetoed on the basis that Bratton is not a British citizen. Instead, he was offered an advisory role on controlling violence, which he gladly accepted.[5]

The UCLA team has since raised several million dollars in venture funding and spun their algorithm out as a private company, which they named PredPol.[6] In December 2012, Pred-Pol made it to England, with a four-month, $200,000 trial taking place in Medway, Kent. In that case, the algorithm was credited with a 6 percent fall in street violence. Similar schemes have now taken place in Greater Manchester, West Yorkshire and the Midlands with similarly promising results.[7] Although some local councillors were worried, believing that predictive analytics would leave rural areas without police cover, or else lead to job cuts, others felt the software was innovative and could bring about a more effective use of resources.[8]

As Malinowski says, predictive policing isn't simply a matter of catching criminals. "What we're trying to do is to be in the right place at the right time, so that when the bad guy shows up he sees the police and is deterred from committing a crime." In the end it all comes back to supermarkets. "We're like a greeter in Walmart," Malinowski says. "It just puts people on notice that you're looking at them."

The Moral Statisticians

The idea of integrating statistics into the world of criminology might seem new. In fact, its roots go back to 19th-century France and to two men named André-Michel Guerry and Adolphe Quetelet. Both Guerry and Quetelet were talented statisticians, who came to the field after first pursuing other careers. In Guerry's case this had been law. For Quetelet, astronomy. Each was profoundly influenced by the work of a man named Isidore Marie Auguste François Xavier Comte—better known as Auguste Comte. Between 1817 and 1823, Comte had worked on a manuscript entitled *Plan of the Scientific Operations Necessary for the Reorganization of Society*. In it he argued that the ideal method for determining how best to run a society would come from studying it in the manner of the natural sciences. In the same way that Isaac Newton could formulate how physical forces might impact upon an object, so did Comte posit that social scientists should be able to discover the universal laws of "social physics" that would predict human behavior.[9]

This idea appealed immensely to Guerry and Quetelet, who had a shared interest in subjects like criminology. At the age of just 26, Guerry had been hired by the French Ministry of Justice to work in a new field called "moral statistics." Quetelet, meanwhile, was enthusiastic about the opportunity to take the mathematical tools of astronomy and apply them to social data. To him:

The possibility of establishing moral statistics, and deducing instructive and useful consequences therefrom, depends entirely on this fundamental fact, that man's free choice disappears, and remains without sensible effect, when the observations are extended over a great number of individuals.[10]

Of benefit to Guerry and Quetelet was the fact that each was living through what can be described as the first "golden age" of Big Data. From 1825, the Ministry of Justice had ordered the creation of the first centralized, national system of crime reporting—to be collected every three months from each region of the country, and which recorded all criminal charges brought before French courts. These broke crimes down into the category of charge, the sex and occupation of the accused, and the eventual outcome in court. Other nationally held data sets included statistics concerning individual wealth (indicated through taxation), levels of entrepreneurship (measured through number of patents filed), the percentage of military troops who could both read and write, immigration and age distribution around the country, and even detailed lists of Parisian prostitutes—ordered by year and place of birth.[11]

During the late 1820s and early 1830s, Guerry and Quetelet worked independently to analyze the available data. One of the first things each remarked upon was the lack of variance that existed in crime from year to year. This had particular relevance in the field of social reform—since reformers had previously focused on redeeming the individual criminal, rather than viewing them as symptoms of a larger problem.[12] Quetelet referred to "the terrifying exactitude with which crimes reproduce themselves" and observed that this consistency carried over even to a granular scale—meaning that the proportion of murders committed by gun, sword, knife, cane, stones, fire, strangulation, drowning, kicks, punches and miscellaneous instruments used for cutting and stabbing remained almost entirely stable on an annual basis. "We know in advance," he proclaimed, "how many individuals will dirty their hands with the blood of others; how many will be forgers; how many poisoners—nearly as well as one can enumerate in advance the births and deaths that must take place." Guerry, too, was struck by "this fixity, this constancy in the reproduction of facts," in which he saw ample evidence that Comte's theories of social physics were correct; that amid the noise of unfiltered data there emitted the dim glow of a signal.

A number of fascinating tidbits emerged from the study of the two scholars. For instance, Quetelet noticed a higher than usual correlation when examining the relationship between suicide by hanging and marriages involving a woman in her twenties and a man in his sixties. Not to be outdone, Guerry also turned his attention to suicide (subdivided by motive and method for ending one's life) and concluded that younger men favored death by pistol, while older males tended toward hanging.

Other relationships proved more complex. Previously, it had been widely thought that poverty was the biggest cause of crime, which meant that wealthier regions of the country would surely have a lower crime rate than poorer ones. In fact, Guerry and Quetelet demonstrated that this was not necessarily the case. While the wealthiest regions of France certainly had lower rates of violent crime than did poorer regions, they also experienced far higher rates of property crime. From this, Guerry was able to suggest that poverty itself was not the cause of property crime. Instead, he pointed to opportunity as the culprit, and argued that in wealthier areas there is more available to steal. Quetelet built on this notion by suggesting the idea of "relative poverty"—meaning that great inequality between poverty and wealth in the same area played a key role in both property and violent crimes. To Quetelet, relative poverty incited people to commit crimes through envy. This was especially true in places where changing economic conditions meant the impoverishment of some, while at the same time allowing others to retain (or even expand) their wealth. Quetelet found less crime in poor areas than in wealthier areas, so long as the people in the poor areas were able to satisfy their basic needs.

Guerry published his findings in a slim 1832 volume called *Essai sur la statistique morale de la France* (*Essay on the Moral Statistics of France*). Quetelet followed with his own *Sur l'homme et le développement de ses facultés* (*On Man and the Development of His Faculties*) three years later. Both works proved immediately sensational: rare instances in which the conclusions of a previously obscure branch of academia truly captures the popular imagination. Guerry and Quetelet were translated into a number of different languages and widely reviewed. The *Westminster Review*—an English magazine founded by Utilitarians John Stuart Mill and Jeremy Bentham—devoted a particularly large amount of space to Guerry's book, which it praised for being of "substantial interest and importance." Charles Darwin read Quetelet's work, as did Fyodor Dostoyevsky (twice), while no less a social reformer than Florence Nightingale based her statistical methods upon his own.[13] Nightingale later gushingly credited Quetelet's findings with "teaching us . . . the laws by which our Moral Progress is to be attained."[14]

In all, Guerry and Quetelet's work showed that human beings were beginning to be understood—not as free-willed, self-determining creatures able to do anything that they wanted, but as beings whose actions were determined by biological and cultural factors.

In other words, they were beings that could be predicted.

The Real Minority Report

In 2002, the Steven Spielberg movie *Minority Report* was released. Starring Tom Cruise and based on a short story by science-fiction author Philip K. Dick, the film tells the story of a futuristic world in which crime has been all but wiped out. This is the result of a specialized "PreCrime" police department, which uses predictions made by three psychics ("precogs") to apprehend potential criminals based on foreknowledge of the crimes they are about to commit.

The advantage of such a PreCrime unit is clear: in the world depicted by *Minority Report*, perpetrators can be arrested and charged as if they had committed a crime, even without the crime in question having to have actually taken place. These forecasts prove so uncannily accurate that at the start of the movie the audience is informed that Washington, D.C.—where the story is set—has remained murder-free for the past six years.

While *Minority Report* is clearly science fiction, like a lot of good sci-fi the world it takes place in is not a million miles away from our own. Even before Sean Malinowski's predictive policing came along, law enforcement officials forecasted on a daily basis by deciding between what are considered "good risks" and "bad risks." Every time a judge sets bail, for instance, he is determining the future likelihood that an individual will return for trial at a certain date. Search warrants are similar predictions that contraband will be found in a particular location. Whenever police officers intervene in domestic violence incidents, their job is to make forecasts about the likely course of that household over time—making an arrest if they feel the future risks are high enough to warrant their doing so. With each of these cases the question of accuracy comes down to both the quality of data being analyzed and the choice of metrics that the decision-making process is based on. With human fallibility being what is, however, this is easier said than done.

Parole hearings represent another example of forecasting. Using information about how a prisoner has behaved while incarcerated, their own plans for their future if released, and usually a psychiatrist's predictions about whether or not they are likely to serve as a danger to the public, parole boards have the option of freeing prisoners prior to the completion of their maximum sentence. According to a 2010 study, though, the single most important factor in determining whether a prisoner is paroled or not may be nothing more scientific than the time of day that their hearing happens to take place. The unwitting participants in this particular study were eight parole judges in Israel. In a situation where entire days are spent reviewing parole applications (each of which lasts for an average of six minutes), the study's author plotted the number of parole requests approved throughout the day. They discovered that parole approval rates peaked at 65 percent after each of the judge's three meal breaks, and steadily declined in the time afterward—eventually hitting zero immediately prior to the next meal.[15] The study suggests that when fatigue and hunger reach a certain point, judges are likely to revert to their default position of denying parole requests. Even though each of the judges would likely place the importance of facts, reason and objectivity over the rumbling of their stomachs, this illustrates the type of problem that rears its head when decision-makers happen to be human.

Richard Berk relies on no such gut instinct. Professor of criminology and statistics at the Wharton School of the University of Pennsylvania, Berk is an expert in the field of computational criminology, a hybrid of criminology, computer science and applied mathematics. For the past decade, he has been working on an algorithm designed to forecast the likelihood of individuals committing violent crimes. In a sly nod to the world of "precogs" envisioned by Philip K. Dick, Berk calls his system "RealCog." "We're not in the world of *Minority Report* yet," he says, "but there's no question that we're heading there." With Berk's RealCog algorithm, he can aid law enforcement officials in making a number of important decisions. "I can help parole boards decide who to release," he says, rattling off the areas he can (and does) regularly advise on. "I can help probation and parole departments decide how best to supervise individuals; I can help judges determine the sentences that are appropriate; I can help departments of social services predict which of the families that are in their area will have children at a high risk of child abuse."

In a previous life, Berk was a sociologist by trade. After receiving his bachelor's in psychology from Yale, followed by a PhD from the Johns Hopkins University, he took a job working as an assistant professor of sociology at Northwestern. He regularly published articles on subjects like the best way to establish a rapport with deviant individuals, and how to bridge the gap between public institutions and people living in poor urban areas. Then he changed tack. "I got interested in statistics as a discipline and pretty much abandoned sociology," he says. "I have not been in a sociology department for decades." A self-described pragmatist, Berk saw the academic work around him producing great insights but doing very little to change the way that things actually worked. When he discovered the field of machine learning, it was a godsend. For the first time he was able to use large datasets, detailing more than 60,000 crimes, along with complex algorithms so that statistical tools could all but replace clinical judgment.

Are You a Darth Vader or a Luke Skywalker?

Berk's algorithm is a black box, meaning that its insides are complex, mysterious and unknowable. What goes in is a dataset of x's, containing information on an individual's background, as well as demographic variables like their age and gender. What comes out is a set of y's, representing the risk they are calculated as posing. The challenge, Berk says, is to find the best formula to allow x to predict y. "When a new person comes through the door, we want to find out whether they are high risk or low risk," he explains. "You enter that person's ID, which is sent out to the various databases. The information in those databases is brought back to the local computer, which figures out what the risk is. That information is then passed along to the decision-maker."

Berk makes no apology for the opacity of his system. "It frees me up," he explains. "I get to try different black boxes and pick the one that forecasts best." What he doesn't care about is causal models. "I make no claims whatsoever that what I'm doing explains *why* it is that individuals fail," he says. "I'm not trying to develop a cause-and-effect rendering of whatever it is that's going on. I just want to forecast accurately."

For this reason, Berk will place more emphasis on what he personally considers to be a strong predictor—such as a prisoner missing work assignments—than he will on full psychological evaluations. If it turns out that liver spots can tell him something about a person's risk of committing future crime, they become a part of his model, with no questions asked. "I don't have to understand why [liver] spots work," Berk says. "If the computer finds things I'm unaware of, I don't care what they are, just so long as they forecast. I'm not trying to explain."

The majority of Berk's metrics for predicting future dangerousness are somewhat more obvious than liver spots. Men are more likely to commit violent crimes than women, while younger men are more likely to behave violently than older men. A man's odds of committing a violent crime are at their highest when he is in his mid-twenties. From there, the chance steadily decreases until the age of 40, after which they plummet to almost zero. It is for this reason that violent crimes committed early in life are more predictive of future crime than those committed later on.

"People assume that if someone murdered, then they will murder in the future," Berk has noted. "But what really matters is what that person did as a young individual. If they committed armed robbery at age 14 that's a good predictor. If they committed the same crime at age 30, that doesn't predict very much."

There is, of course, the question of potential errors. Berk claims that his forecasting system can predict with 75 percent accuracy whether a person released on parole will be involved in a homicide at some point in the future. That is certainly an impressive number, but one that still means he will be wrong as often as one in every four times. "No matter how good this forecasting is going to be we are always going to make mistakes," Berk acknowledges. "Everyone appreciates this, although that doesn't make it any less painful." The margin for error in crime prediction is something of a theme in the film *Minority Report*, where the "minority report" alluded to in the title refers to the suppressed information that the three precogs used to predict crimes occasionally disagree on predictions. "Are you saying I've [arrested] innocent people?" asks Tom Cruise's police chief when he discovers that this vital piece of information has been kept from him. "I'm saying that every so often, those accused of a PreCrime might, *just might*, have an alternate future," comes the answer. It is easy to see why news of such minority reports might have been kept from the public. In order for PreCrime to function, there can be no

suggestion of fallibility. Who wants a justice system that in-
stills doubt, no matter how effective it might be?

To Berk, mistakes come in one of two different forms: false
positives and false negatives. A false positive (which he refers to
as a "Luke Skywalker") is a person incorrectly identified as a
high-risk individual. A false negative (a "Darth Vader") is a high-
risk individual who is not recognized as such. This, Berk says, is
a political question rather than a statistical one. Is it worse to
falsely accuse a Luke Skywalker, or fail to find a Darth Vader?
"In theory, false positives and false negatives are treated as
equally serious errors," he says. "But in practice, this turns out
not to be true. If you work with criminal justice officials, or talk
to stakeholders or citizens, some mistakes are worse than others.
In general it is worse to fail to identify a high-risk individual,
than it is to falsely represent someone as if they were a high-risk
individual." That is therefore the way that Berk's algorithm is
weighted. In many criminal justice settings—particularly when
it comes to violent crime—decision-makers are likely to accept
weaker evidence if this means avoiding letting a potential Darth
Vader slip through the cracks. The price that you pay is that
more Luke Skywalkers will be falsely accused.

Soon, Berk says, the availability of data sources will expand
even further. Rather than just relying on the official records
routinely available in criminal-justice files, Berk and his colleagues will be able to use custom data sources to predict
future criminality. GPS ankle bracelets, for instance, can
examine how people spend their free time—and algorithms
can then compare this to a database of criminals to see whether
a person happens to share similar pastimes with an Al Capone
or a Ted Bundy. "Are they at home watching TV, or are they

spending [their free time] on a particular street corner, which historically has been used for drug transactions?" Berk asked in a 2012 talk for Chicago Ideas Week. "Knowing whether someone is out at 2 A.M. on a Saturday morning versus 10 A.M. will, we think, help us forecast better."

Given time it might even prove possible to begin building up profiles of people before they commit a crime in the first place: one step closer to the authoritarian world of PreCrime imagined by Philip K. Dick. As fraught with legal and ethical dilemmas as this is, a person who grows up in a high-crime area, has a family history of drug abuse, or perhaps a sibling or parent already in prison, could be identified by Berk's algorithm—despite not yet having an arrest history.

Another area in which such technology will likely be used is as part of the school system. "Schools want to know whether the students they have are going to create problems," Berk says. "There's some initial work being done using school data for school kids who have no criminal record, to determine which are high risk for dropping out of school, for being truant, for getting in fights, for vandalism, and so on." In 2013, he was working with children, aged between 8 and 10, to use predictive modeling to establish how likely it is that they may commit a felony later in life. For now, however, realizing this technology remains the stuff of the future. As Berk himself acknowledges, "That's just a gleam in our eye. We're just starting to do that."

Delete All the Lawyers

There is a line in Shakespeare's *Henry VI* that is likely a favorite of anyone who has ever had adverse dealings with those in the legal profession. Adding his two cents to a plan to stage a social revolt in England, the character of Dick the Butcher pipes up with a suggestion he sees as all but guaranteeing utopia. "The first thing we do," he says, "let's kill all the lawyers."

Approaching 500 years after Shakespeare's play was first performed, lawyers might not yet be dead, but The Formula may be rendering an increasing number of them irrelevant. Consider the area of legal discovery, for instance. Legal discovery refers to the pretrial phase of a lawsuit, in which each party obtains and sorts through the material it requires that may lead to admissible evidence in court. In previous years, discovery was mainly carried out by junior lawyers, dispatched by law firms to comb through large quantities of possible evidence by hand. This task was both good for law firms and bad for clients, who inevitably found themselves on the receiving end of costly fees. In 1978, five television studios became entangled in an antitrust lawsuit filed against broadcasting giant CBS. To examine the 6 million documents deemed to be relevant to the case, the five television studios hired a team of lawyers who worked for a period of several months. When their bill eventually came in, it was for an amount in excess of $2.2 million—close to $8 million in today's money.[16]

Thanks to advances in artificial intelligence, today discovery can be carried out using data-mining "e-discovery" tools, in addition to machine-learning processes such as predictive coding. Predictive coding allows for human lawyers to manually review a small percentage of the available documents, and in doing so to "teach" the computer to distinguish between relevant and irrelevant information. Algorithms can then process the bulk of the information in less than a third of the time it would take for even the most competent of human teams to carry out the same task. Such systems have shown a repeated ability to outperform both junior lawyers and paralegals in terms of precision. After all, computers don't get headaches.

Often e-discovery requires nothing more complex than basic indexing or the imposing of simple legal classification. It can be used to go further, though. Algorithms can extract relevant concepts (such as pulling out all documents pertaining to social protest in the Middle East) and are becoming increasingly adept at searching for specific ideas, regardless of the wording used. Algorithms can even search for the *absence* of particular terms, or look for the kind of underlying patterns that would likely have eluded the attention of human lawyers. Better yet, not only can this work be carried out faster than it would take human lawyers—but for a fraction of the cost as well. One of the world's leading e-discovery firms, Palo Alto's Blackstone Electronic Discovery, is regularly able to analyze 1.5 million documents for a cost of less than $100,000.

Located in the heart of Silicon Valley, Blackstone boasts a client list that includes Netflix, Adobe and the United States Department of Justice. In 2012, the company worked on the *Apple v. Samsung* patent case, described by *Fortune* magazine as the "(Patent) trial of the century."[17] Blackstone's founder is an MBA named John Kelly, who started the company in 2003 in response to what he saw as the obvious direction the legal profession was headed in. "The amount of data we have to deal with today is absolutely exploding," Kelly says. "Twenty years ago a typical case might have involved ten boxes of hard copy. Today it's easy to pull 100GB of soft copy just from local servers, desktops and smartphones. That's the equivalent of between 1 and 2 million pages right there."

Part of the explanation for the exponential increase in data comes down to the ease with which information can now be stored. Rather than giant, space-consuming filing cabinets of physical documents, modern companies increasingly store their data in the form of digital files. In a world of cloud-based storage solutions there is literally no excuse for throwing anything away. In the *Apple v. Samsung* case, Samsung found itself verbally chastised by the presiding judge after admitting that all corporate e-mails carried an expiry date, which caused them to be automatically deleted every two weeks. As *Fast Company* pointed out, "Samsung lost [the case] anyway, but this [infraction] . . . might have sealed the outcome."[18]

Where previously the field of discovery was about finding enough data to build a case, now the e-discovery process focuses instead on how much information can be safely *ignored*. As Kelly points out, "In the digital world, it's more about figuring out how to limit the scope of our investigations." This is another application in which the algorithms come into their own. Kelly acknowledges that the efficiency with which this limiting process can be carried out doesn't always endear his company to those working in more traditional law firms. "Some firms might see a case and think of it as a $5 million opportunity," he says. "A company like Blackstone then comes in and for $100,000 can take the number of relevant documents down to just 500 e-mails in the blink of an eye."

If there's one area Kelly admits to worrying about it is for the new generation of junior lawyers, whose livelihood is being threatened by automation. "One of the questions our work provokes is what happens to that cadre of folks just out of law school, who don't have clients yet, who aren't rainmakers— what are they going to do?" Kelly says, with a twinge of genuine pain in his voice. "In the old days there was tons of stuff around for them. It might not always have been exciting work, but at least it was available. Now guys like us can do a lot of that work just by using the right algorithm."

Divorce by Algorithm

Business-management guru Clayton Christensen identifies two types of new technology: "sustaining" and "disruptive" innovations.[19] A sustaining technology is something that supports or enhances the way a business or market already operates. A disruptive technology, on the other hand, fundamentally alters the way in which a particular sector functions. An example of the former might be something like the advent of computerized accounting systems, while the arrival of digital cameras (which famously led to the downfall of Kodak) represents the latter. Tools like e-discovery algorithms are *disruptors*. But they're also far from the exception to the rule when it comes to the many ways in which the legal profession is being irreversibly altered by the arrival of The Formula.

In his classic book, *The Selfish Gene*, evolutionary biologist Richard Dawkins describes the way in which legal cases have "evolved" to become as inefficient as possible—thereby enabling lawyers, working with one another in "elaborately coded cooperation," to jointly milk their clients' bank accounts for the longest amount of time possible. In its own way this is an algorithm in itself, albeit one that is diametrically opposed to a computer-based algorithm designed to produce efficient results in as few steps as possible.[20]

With the unbalanced weighting of the legal system in favor of those practicing it,[21] it is no surprise that many lawyers criticize the use of disruptive technologies in law, worried about the detrimental effects that it is likely to have on their earning power. A large number of these attack what is viewed as the "commoditization" of law: unfavorably comparing "routinized" legal work to the kind of "bespoke" work you would receive from a human lawyer. (Think about the difference between off-the-rack and hand-tailored clothing in both quality and price.)

But while this criticism makes sense if you are a lawyer carrying out what you feel to be bespoke work, it also heavily downplays the number of legal tasks that a bot can perform as well as, if not better than, a person. One such area that The Formula is revolutionizing is the process of contract drafting, thanks to the rise of automated document assembly systems like the snappily named LegalZoom. Founded by two corporate-law refugees in 2001, LegalZoom has since served more than 2 million customers and in the process become a better-known brand name within the United States than any other law firm. Charging as little as $69 for wills and $99 for articles of incorporation, LegalZoom uses algorithms for its high-volume, low-cost business of providing basic consumer and business documents: doing for the legal profession what Craigslist did for the newspaper industry's profitable classified ad business.[22] Another area is trademark analysis, with the Finnish start-up Onomatics having created an algorithm capable of generating instant reports showing how far apart two different trade-

marks might be: an area notorious for its high level of subjective stickiness.

A similar technology is Wevorce, a Mountain View, California, start-up, located several miles from Google's corporate
headquarters. If Internet dating's frictionless approach to
coupling—discussed last chapter—promises to take the pain
out of love, then Wevorce makes the same promise about
divorce: offering a divorce service mediated by algorithm. Not
only does Wevorce provide a standardized service, based on a
system that identifies 18 different archetypal patterns in
divorcing couples—but it can even advise on what stage in the
grieving process a user's former partner is likely to be experiencing at any given moment. By asking couples to think rationally (even computationally) about separation, Wevorce claims
that it can "[change] divorce for the better."[23] "Because the
software keeps the process structured, it's less scary for divorcing couples and more efficient for attorneys, which leads to
overall lower costs," says CEO Michelle Crosby.[24]

The Invisible Law Enforcer

Many technology commentators have remarked on the major
shift that has accompanied our changing understanding of the
term "transparency" over the past 40 years. In the early days of
personal computing, a computer that was transparent meant a
computer on which one could "lift the lid" and tinker with its
inner workings. Jump forward to the present day and transparency instead denotes something entirely more opaque: namely
that we can make something work without understanding *how*
it works. To quote MIT psychoanalyst Sherry Turkle, as we

have moved from a modernist culture of calculation to a post-modern one of simulation, computers increasingly ask to be taken at "interface value."[25] This idea of interface value was perfectly encapsulated during a 2011 interview with Alex Kipman, one of Microsoft's engineers on the Kinect motion-sensing device. Speaking to a reporter from the *New York Times*, Kipman proudly explained that increasingly we are headed toward "a world where technology more fundamentally understands you, so you don't have to understand it."[26]

It is into this frame that "Ambient Law" enters.[27] Ambient Law refers to the idea that instead of requiring lawyers to call attention to items of legal significance around us, laws can be both embedded within and enforced by our devices and environment. Ambient Law is a vision of the future in which autonomic smart environments take an unprecedented number of decisions for and about us on a constant, real-time basis. Autonomic computing's central metaphor is that of the human body's central nervous system. In the same way that the body regulates temperature, breathing and heart rate, without us having to be consciously aware of what is happening, so too is the dream of autonomic computing for algorithms to self-manage, self-configure and self-optimize—without the need for physical or mental input on the part of users.

One example of Ambient Law might be the "smart office," which continuously monitors its own internal temperature and compares these levels to those stipulated by health and safety regulations. In the event that a specified legal limit is exceeded, an alarm could be programmed to sound. Another usage of Ambient Law is the car that refuses to be driven by individuals with an excessive level of alcohol in their bloodstream. A number of different car manufacturers have developed similar technology in recent years. In a system developed by Japanese carmaker Nissan, would-be motorists are monitored from the moment they get behind the wheel. An alcohol odor sensor is used to check their breath, another sensor tests for alcohol in the sweat of their palm as they touch the gear stick, and a miniature dashboard camera monitors their face and eye movements—looking for increased blinking to indicate drowsiness, or a drooping mouth to suggest yawning. Based on an averaging of all of these biometrics, the car's in-built algorithms then decide whether or not an individual is safe to drive. If the answer is negative, the car's transmission locks, a "drunk-driving" voice alert sounds over the car's satellite navigation system, and the driver's seat belt tightens around them to provide (in the words of a Nissan spokesperson) a "mild jolt" designed to snap them out of their stupor.[28]

The Politics of Public Space

These technologies unnerve some people because of what they suggest about algorithms' new role as moral decision-makers. One only has to look at the hostile reaction afforded Apple when it started censoring "objectionable" content in its App Store, to see that many computer users view morality and technology as two unrelated subjects. This is an understandable reaction, but one that also shows a lack of awareness about the historical role of "technology." If science aims for a better understanding of the world, then technology (and, more accurately, technologists) has always sought to change it. The result is a discipline that is inextricably tied in with a sense of morality, regardless of how much certain individuals might try to deny it. In this way, technology is a lot like law, with both designed as man-made forces for regulating human behavior.

In a 1980 essay entitled "Do Artifacts Have Politics?," the sociologist Langdon Winner singled out several of the bridges over the parkways on Long Island, New York.[29] Many of these bridges, Winner observed, were extraordinarily low, with as little as nine feet of clearance at the curb. Although the majority of people seeing them would be unlikely to attach any special meaning to their design, they were actually an embodiment of the social and racial prejudice of designer Robert Moses, who was responsible for building many of the roads, parks, bridges and other public works in New York between the 1920s and 1970s. With the low bridges, Moses's intention was to allow only whites of "upper" and "comfortable middle" classes access to the public park, since these were the only demographics able to afford cars. Because poorer individuals, which included many blacks, relied on taller public buses they were denied access to the park, since the buses were unable to handle the low overpasses and were forced to find alternative routes. In other words, Moses built bias (and a skewed sense of morality) into his designs. As New York town planner Lee Koppleman later recalled, "The old son of a gun . . . made sure that buses would never be able to use his goddamned parkways."[30]

While the neo-libertarian Google might be a million miles from Moses's attitudinal bias, it is difficult not to look at the company's plans to use data-mining algorithms to personalize maps and see (perhaps unintentional) strains of the same stuffy conservatism. Over the past decade, Google Maps has become a ubiquitous part of many people's lives, vital to how we move from one place to another on a daily basis. As journalist Tom Chivers wrote in the *Daily Telegraph*, "Of all of the search giant's many tentacles reaching octopus-like into every area of our existence, Maps, together with its partner Google Earth and their various offspring, can probably claim to be the one that has changed our day-to-day life the most."[31] In 2011, while speaking to the website *TechCrunch*, Daniel Graf, the director of Google Maps for mobile, asked rhetorically, "If *you* look at a map and if *I* look at a map [my emphasis], should it always be the same for you and me? I'm not sure about that, because I go to different places than you do."[32] The result of this insight was that from 2013 onward, Google Maps began incorporating user information to direct users toward those places most likely to be home to like-minded individuals, or subjects that they have previously expressed an interest in. "In the past, such a notion would have been unbelievable," Google crowed in promotional literature. "[A] map was just a map, and you got the same one for New York City, whether you were searching for the Empire State Building or the coffee shop down the street. What if, instead, you had a map that's unique to you, always adapting to the task you want to perform right this minute?"

But while this might be helpful in some senses, its intrinsic "filter bubble" effect may also result in users experiencing less of the serendipitous discovery than they would by using a traditional map. Like the algorithmic matching of a dating site, only those people and places determined on your behalf as suitable or desirable will show up.[33] As such, while applying The Formula to the field of cartography might be a logical step for Google, it is potentially troubling. Stopping people of a lower economic status from seeing the houses and shops catering to those of higher economic means—or those of one religion from seeing on their maps the places of worship belonging to those of another—might initially seem a viable means of reducing conflict, but it would do nothing in the long term to promote tolerance, understanding or an evening of the playing field.

This situation might only be further exacerbated were certain algorithms, like the Nara recommender system I described in Chapter 1, to be implemented incorrectly. By looking not at where an individual currently is, but where advertisers would eventually like them to be, people reliant on algorithms for direction could be channeled down certain routes—like actors playing along to a script.

Your Line, My Line

It was the French philosopher and anthropologist Bruno Latour—picking up from where sociologist Langdon Winner left off—who first put forward the notion of technological "scripts."[34] In the same way that a film script or stage play prescribes the actions of its performers, so too did Latour argue that technology can serve to modify the behavior of its users by demanding to be dealt with in a certain way.[35] For instance, the disposability of a plastic coffee cup, which begins to disintegrate after only several uses, will encourage people to throw it away. A set of heavy weights attached to hotel keys similarly makes it more likely that they will be returned to the reception desk, since the weight will make the keys cumbersome to carry around.

Less subtle might be the springs attached to a door that dictate the speed at which people should enter a building—or the concrete speed bumps that prompt drivers to drive slowly, or else risk damaging their shock absorbers. Less subtle still would be the type of aforementioned Ambient Law that ensures that a vehicle will not start because its driver is inebriated, or the office building that not only sounds an alarm but also turns off workers' computer screens because a certain heat threshold has been reached, and they should exit for their own safety.

As with Moses's low-hanging bridges, such scripts can be purposely inscribed by designers. By doing this, designers delegate specific responsibilities to the objects they create, and these can be used to influence user behavior—whether that be encouraging them to conform to particular social norms or forcing them into obeying certain laws.[36] Because they serve as an "added extra" on top of the basic functionality of an object or device, scripts pose a number of ethical questions. What, for example, is the specific responsibility of the technology designer who serves as the inscriber of scripts? If laws or rules are an effort to moralize other people, does this differ from attempts to moralize technology? Can we quantify in any real sense the difference between a rule that asks that we not waste water in the shower and the use of a water-saving showerhead technology that ensures that we do not?

In their book *Nudge: Improving Decisions about Health, Wealth, and Happiness,* authors Richard Thaler and Cass Sunstein recount the story of a fake housefly placed in each of the urinals at Schiphol Airport in Amsterdam. By giving urinating men something to aim at, spillage was reduced by a whole 80 percent.[37] While few would likely decry the kind of soft paternalism designed to keep public toilets clean, what about the harder paternalism of a car that forcibly brakes to stop a person breaking the speed limit? To what degree can actions be considered moral or law-abiding if the person carrying them out has no choice but to do so? And when particular actions are inscribed by designers (or, in the case of The Formula, computer scientists), who has the right to implement and enforce them?

While it would be a brave (and likely misguided) person who would step up and defend a drunk driver's right to drive purely on democratic grounds, the question of the degree to which behavior should be rightfully limited or regulated is one central to moral philosophy. In some situations it may appear to be morally justified. In others, it could just as easily raise associations with the kind of totalitarian technocracy predicted in George Orwell's *Nineteen Eighty-Four.*

Unsurprisingly, there is a high level of disagreement about where the line in the sand should be drawn. Roger Brownsword, a legal scholar who has written extensively on the topic of technological regulation and the law, argues that the autonomy that underpins human rights means that a person should have the option of either obeying or disobeying a particular rule.[38] At the other end of the spectrum is Professor Sarah Conly, whose boldly titled book, *Against Autonomy*, advocates "[saving] people from themselves" by banning anything that might prove physically or psychologically detrimental to their well-being. These include (but are by no means limited to) cigarettes, trans fats, excessively sized meals, the ability to rack up large amounts of debt, and the spending of too much of one's paycheck without first making the proper saving provisions. "Sometimes no amount of public education can get someone to realize, in a sufficiently vivid sense, the potential dangers of his course of behavior," Conly writes. "If public education were effective, we would have no new smokers, but we do." Needless to say, in Conly's world of hard paternalism there are more speed bumps than there are plastic coffee cups.[39]

The Prius and the Learning Tree

On the surface, the idea that we should be able to enforce laws by algorithm makes a lot of sense. Since legal reasoning is logical by nature, and logical operations can be automated by a computer, couldn't codifying the legal process help make it more efficient than it already is? In this scenario, deciding legal cases would simply be a matter of entering the facts of a particular case, applying the rules to the facts, and ultimately determining the "correct" answer.

In his work, American scholar Lawrence Lessig identifies law and computer code as two sides of the same coin. Lessig refers to the laws created by Congress in Washington, D.C., as "East Coast code" and the laws that govern computer programs as "West Coast code," in reference to the location of Silicon Valley.[40] Once a law of either type is created, Lessig argues, it becomes virtual in the sense that from this point forward it has an existence independent of its original creator. Lessig was hardly the first person to explore this similarity. Three hundred years before Lessig's birth, the great mathematician and co-inventor of calculus, Gottfried Leibniz, speculated that legal liability could be determined using calculation. Toward the end of the 19th century another larger group of legal scholars formed the so-called jurimetrics movement, which argued that the "ideal system of law should draw its postulates and its legis-lative justification from science."[41]

While both Leibniz and the jurimetrics were misguided in their imagining of the legal system as a series of static natural laws, their dream was—at its root—an honest one: based on the idea that science could be used to make the law more objective. Objectivity in a legal setting means fairness and impartiality. The person who fails to act objectively has allowed self-interest or prejudice to cloud their judgment. By attempting to turn legal reasoning into a system that would interpret rules the same way every time, the belief was that a consistency could be found to rival that which is seen in the hard sciences.

The problem with the jurimetrics' approach to law was challenged most effectively by an experiment carried out in 2013—designed to examine the challenges of turning even the most straightforward of laws into an algorithm. For the study, 52 computer programmers were assembled and split into two groups. Each group was tasked with creating an algorithm that would issue speeding tickets to the driver of a car whenever it broke the speed limit. Both groups were pro-vided with two datasets: the legal speed limit along a particu-lar route, and the information about the speed of a particular vehicle (a Toyota Prius) traveling that route on a previous occasion, collected by using an on-board computer. The data showed that the Prius rarely exceeded the speed limit, and on those occasions that it did, did so only briefly and by a moder-ate degree. To make things more morally ambiguous, these violations occurred at times during the journey when the Prius was set to cruise control. The journey was ultimately completed safely and without incident.

The first group of computer programmers was asked to write their algorithm so that it conformed to "the letter of the law." The second was asked to meet "the intent of the law." Unsurprisingly, both came to very different conclusions. The "intent of the law" group issued between zero and 1.5 tickets to the driver for the journey. The "letter of the law" group, on the other hand, issued a far more draconian 498.3 tickets. The astonishing disparity between the two groups came down to two principal factors. Where the "intent of the law" group allowed a small amount of leeway when crossing the speed limit, the "letter of the law" group did not. The "letter of the law" group also treated each separate sample above the speed limit as a new offense, thereby allowing a continuous stream of tickets to be issued in a manner not possible using single speed cameras.[42]

Rules and Standards

This raises the question of "rules" versus "standards." Broadly speaking, individual laws can be divided up into these two distinct camps, each of which exists on opposite ends of the legal spectrum. To illustrate the difference between a rule and a standard, consider two potential laws both designed to crack down on unsafe driving. A rule might state, "No vehicle shall drive faster than 65 miles per hour." A standard, on the other hand, may be articulated as "No one shall drive at unsafe speeds." The subjectivity of standards means that they require human discretion to implement, while rules exist as hard-line binary decisions, with very little in the way of flexibility.

Computers are constantly getting better at dealing with the

kind of contextual problems required to operate in the real world. For example, an algorithm could be taught to abandon enforced minimum speed limits on major roads in the event that traffic or weather conditions make adhering to them impossible. Algorithms used for processing speed camera information have already shown themselves capable of picking out people who are new to a certain area. Drivers who are unfamiliar with a place might find themselves let off if they are marginally in excess of the speed limit, while locals spotted regularly along a particular route may find themselves subject to harsher treatment. However, both of these exceptions take the form of preprogrammed rules, as opposed to evidence that computers are good at dealing with matters of ambiguity.

This technological drive toward more objective "rules" is one that has played out over the past several centuries. For instance, in his book *Medicine and the Reign of Technology*, Stanley Joel Reiser observes how the invention of the stethoscope

> helped to create the objective physician, who could move away from involvement with the patient's experiences and sensations, to a more detached relation, less with the patient but more with the sounds from within the body.[43]

Similar sentiments are echoed by sociologist Joseph Gusfield in *The Culture of Public Problems: Drinking-Driving and the Symbolic Order*, in which he argues that the rise of law enforcement technologies stems from a desire for objectivity, centered around that which is quantifiable. To illustrate his point, Gusfield looks at the effect that the introduction of the

Breathalyzer in the 1950s had on the previously subjective observations law enforcement officials had relied on to determine whether or not a person was "under the influence." As Gusfield writes:

> [Prior to such tests, there was] a morass of uncorroborated reports, individual judgments, and criteria difficult to apply to each case in the same manner. Both at law and in the research "laboratory," the technology of the blood level sample and the Breathalyzer meant a definitive and easily validated measure of the amount of alcohol in the blood and, consequently, an accentuated law enforcement and a higher expectancy of convictions.[44]

In other words, the arrival of the Breathalyzer turned a person's ability to drive after several drinks from abstract "standard" into concrete "rule" in the eyes of the law. This issue will become even more pressing as the rise of Ambient Law continues—with technologies not only having the power to regulate behavior but to dictate it as well, sometimes by barring particular courses of action from being taken.

Several years ago, Google announced that it was working on a fleet of self-driving cars, in which algorithms would be used for everything from planning the most efficient journey routes, to changing lanes on the motorway by determining the smoothest path combining trajectory, speed and safe distance from nearby obstacles. At the time of writing, these cars have completed upward of 300,000 miles of test drives in a wide range of conditions, without any reported accidents—leading to the suggestion that a person is safer in a car driven by an algorithm than they are in one driven by a human.[45] Since cars driven by an algorithm already conform to a series of preprogrammed rules, it is understandable why specific laws would become just more to add to the collection. These could lead to a number of ethical challenges, however. As a straightforward example, what would happen if the passenger in the car needed to reach a hospital as a matter of urgency—and that this meant breaking the speed limit on a largely empty stretch of road? It is one thing if the driver/passenger was ticketed at a later date thanks to the car's built-in speed tracker. But what if the self-driving car, bound by fixed Ambient Laws, refused to break the regulated speed limit under any conditions?

You might not even have to wait for the arrival of self-driving cars for such a scenario to become reality. In 2013, British newspapers reported on road-safety measures being drawn up by EC officials in Brussels that would see all new cars fitted with "Intelligent Speed Adaptation" measures similar to those already installed in many heavy-goods vehicles and buses. Using satellite feeds, or cameras designed to automatically detect and read road signs, vehicles could be forced to conform to speed limits. Attempts to exceed them would result in the deployment of the car's brakes.[46]

One Man's Infrastructure Is Another Man's Difficulty

The move toward more rule-based approaches to law will only continue to deepen as more and more laws are written with algorithms and software in mind. The fundamental problem, however, is that while rules and standards do exist as opposite abstract poles in terms of legal reasoning, they are exactly that: abstract concepts. The majority of rules carry a degree of "standard-ness," while many standards will be "rule-ish" in one way or another. Driving at a maximum of 60 miles per hour along a particular stretch of road might be a rule (as opposed to the standard "don't drive at unsafe speeds"), but the fact that we might not receive a speeding ticket for driving at, say, 62 miles per hour suggests that these are not hard-and-fast in their rule-ishness. While standards might be open to too much subjectivity, rules may also not be desirable on account of their lack of flexibility in achieving broader social goals.

As an illustration, consider the idea of a hypothetical sign at the entrance of a public park that states, "No vehicles are allowed in this park." Such a law could be enforced through the use of CCTV cameras, equipped with algorithms designed to recognize moving objects that do not conform to the shape of a human. In the hard paternalistic "script" version, cameras might be positioned at park entrances and linked to gates that are automatically opened only when the algorithm is satisfied that all members of a party are on foot. In the "softer" paternalistic version, cameras may be positioned all over the park and could identify the driver of any vehicles seen inside park boundaries by matching their face to a database of ID images and then issuing an automated fine, which is sent directly to the offender's home address.

This might sound fair—particularly if the park is one that has previously experienced problems with people driving their cars through it. But would such a rule also apply to a bicycle? Deductive reasoning may well state that since a bicycle is a vehicle also, any law that states that no vehicles should be allowed in a park must also apply to a bicycle. However, is this the intention of the law, or is the law in this case designed to stop motor vehicles entering a park since they will create noise and pollution? Does such a rule also mean barring the entrance of an ambulance that needs to enter the park to save a person's life? And if it does not, does this mean that algorithmic laws would have to be written in such a way that they would not apply to certain classes of citizen?

One more obvious challenge in a world that promises "technology [that] more fundamentally understands you, so you don't have to understand it" is that people might break laws they are not even aware of. In many parts of the world, it is against the law to be drunk in public. With CCTV cameras increasingly equipped with the kind of facial and even gait recognition technology (i.e., analyzing the way you walk) that might allow algorithms to predict whether or not a person is drunk, could an individual be ticketed for staggering home on foot after several drinks at the pub? Much the same is true of cycling at night without the presence of amber reflectors on bike pedals, which is illegal in the UK—or jaywalking, which is legal in the UK but illegal in the United States, Europe and Australia.[47] In both of these cases, facial-recognition technology could be used to identify individuals and charge them. As a leading CCTV industry representative previously explained in an article for trade magazine *CCTV Today*, "Recognizing aberrant behavior is for a scientist a matter of grouping expected behavior and writing an algorithm that recognizes any deviation from the 'normal.'"[48]

I should hardly have to point out what is wrong with this statement. "Normal" behavior is not an objective measure, but rather a social construct—with all of the human bias that suggests.

Not all uses of algorithmic surveillance are immediately prejudiced, of course. For example, researchers have developed an algorithm for spotting potential suicide jumpers on the London Underground, by watching for individuals who wait on the platform for at least ten minutes and miss several available trains during that time.[49] If this proves to be the case, the algorithm triggers an alarm. Another potential application helps spot fights breaking out on the street by identifying individuals whose legs and arms are moving back and forth in rapid motion, suggesting punches and kicks being thrown. Things become more questionable, however, when an algorithm might be used to alert authorities of a gathering crowd in a location where none is expected. Similarly, in countries with overtly discriminatory laws, algorithms could become a means by which to intimidate and marginalize members of the public. In Russia, where the gay community has been targeted by retrograde antigay laws, algorithmic surveillance may be a means of identifying same-sex couples exhibiting affectionate behavior.

In such a scenario, algorithms would function in the opposite way to that which I described in Chapter 1. Where companies like Quantcast and Amazon prize "aberrant" behavior on the basis that it gives them insights into the unique behavior of individual users, algorithms could instead become a way of ensuring that people conform to similar behavior—or else. As the American sociologist Susan Star once phrased it, one person's infrastructure is another's difficulty.[50]

All of these are (somewhat alarmingly) examples of what would happen if law enforcement algorithms got it right: upholding the rules with the kind of steely determination that would put even fictitious hard-nosed lawman Judge Dredd to shame. What would happen if they got things wrong, on the other hand, is potentially even more frightening. . . .

The Deadbeat Dad Algorithm

On April 5, 2011, 41-year-old John Gass received a letter from the Massachusetts Registry of Motor Vehicles. The letter informed Gass that his driver's license had been revoked and that he should stop driving, effective immediately. The only problem was that, as a conscientious driver who had not received so much as a traffic violation in years, Gass had no idea why it had been sent. After several frantic phone calls, followed up by a hearing with Registry officials, he learned the reason: his image had been automatically flagged by a facial-recognition algorithm designed to scan through a database of millions of state driver's licenses looking for potential criminal false identities. The algorithm had determined that Gass looked sufficiently like another Massachusetts driver that foul play was likely involved—and the automated letter from the Registry of Motor Vehicles was the end result. The RMV itself was unsympathetic, claiming that it was the accused individual's "burden" to clear his or her name in the event of any mistakes, and arguing that the pros of protecting the public far outweighed the inconvenience to the wrongly targeted few.[51]

John Gass is hardly alone in being a victim of algorithms gone awry. In 2007, a glitch in the California Department of Health Services' new automated computer system terminated the benefits of thousands of low-income seniors and people with disabilities. Without their premiums paid, Medicare canceled those citizens' health care coverage.[52] Where the previous system had notified people considered no longer eligible for benefits by sending them a letter through the mail, the replacement CalWIN software was designed to cut them off without notice, unless they manually logged in and prevented this from happening. As a result, a large number of those whose premiums were discontinued did not realize what had happened until they started receiving expensive medical bills through the mail. Even then, many lacked the necessary English skills to be able to navigate the online health care system to find out what had gone wrong.[53]

Similar faults have seen voters expunged from electoral rolls without notice, small businesses labeled as ineligible for government contracts, and individuals mistakenly identified as "deadbeat" parents. In a notable example of the latter, 56-year-old mechanic Walter Vollmer was incorrectly targeted by the Federal Parent Locator Service and issued a child-support bill for the sum of $206,000. Vollmer's wife of 32 years became suicidal in the aftermath, believing that her husband had been leading a secret life for much of their marriage.[54]

Equally alarming is the possibility that an algorithm may falsely profile an individual as a terrorist: a fate that befalls roughly 1,500 unlucky airline travelers each week.[55] Those fingered in the past as the result of data-matching errors include former Army majors, a four-year-old boy, and an American Airlines pilot—who was detained 80 times over the course of a single year.

Many of these problems are the result of the new roles algorithms play in law enforcement. As slashed budgets lead to increased staff cuts, automated systems have moved from simple administrative tools to become primary decision-makers.

In a number of cases, the problem is about more than simply finding the right algorithm for the job, but about the problematic nature of believing that any and all tasks can be automated to begin with. Take the subject of using data-mining to uncover terrorist plots, for instance. With such attacks statistically rare and not conforming to well-defined profiles in the way that, for example, Amazon purchases do, individual travelers end up surrendering large amounts of personal privacy to data-mining algorithms, with little but false alarms to show for it. As renowned computer security expert Bruce Schneier has noted:

> Finding terrorism plots . . . is a needle-in-a-haystack problem, and throwing more hay on the pile doesn't make that problem any easier. We'd be far better off putting people in charge of investigating potential plots and letting them direct the computers, instead of putting the computers in charge and letting them decide who should be investigated.[56]

While it is clear why such emotive subjects would be considered ripe for The Formula, the central problem once again comes down to the spectral promise of algorithmic objectivity. "We are all so scared of human bias and inconsistency," says Danielle Citron, professor of law at the University of Maryland. "At the same time, we are overconfident about what it is that computers can do." The mistake, Citron suggests, is that we "trust algorithms, because we think of them as objective, whereas the reality is that humans craft those algorithms and can embed in them all sorts of biases and perspectives." To put it another way, a computer algorithm might be unbiased in its execution, but, as noted, this does not mean that there is not bias encoded within it. What the speed limit algorithm experiment mentioned earlier in this chapter shows more than anything is the degree to which assumptions are built into the code that computer programmers write, even when those problems being solved might be relatively mechanical in nature. As technology historian Melvin Kranzberg's first law of technology states: "Technology is neither good nor bad—nor is it neutral."

Implicit or explicit biases might be the work of one or two human programmers, or else come down to technological difficulties. For example, algorithms used in facial recognition technology have in the past shown higher identification rates for men than for women, and for individuals of non-white origin than for whites. An algorithm might not target an African-American male for reasons of overt prejudice, but the fact that it is more likely to do this than it is to target a white female means that the end result is no different.[57] Biases can also come in the abstract patterns hidden within a dataset's chaos of correlations.

Consider the story of African-American Harvard University PhD Latanya Sweeney, for instance. Searching on Google one day, Sweeney was shocked to notice that her search results were accompanied by adverts asking, "Have you ever been arrested?" These ads did not appear for her white colleagues. Sweeney began a study that ultimately demonstrated that the machine-learning tools behind Google's search were being inadvertently racist, by linking names more commonly given to black people to ads relating to arrest records.[58] A similar revelation is the fact that Google Play's recommender system suggests users who download

Grindr, a location-based social-networking tool for gay men, also download a sex-offender location-tracking app. In both of these cases, are we to assume that the algorithm has made an error, or that they are revealing inbuilt prejudice on the part of their makers? Or, as is more likely, are they revealing distasteful large-scale cultural associations between—in the former case—black people and criminal behavior and—in the latter—homosexuality and predatory behavior?[59] Regardless of the reason, no matter how reprehensible these codified links might be, they demonstrate another part of algorithmic culture. A single human showing explicit bias can only ever affect a finite number of people. An algorithm, on the other hand, has the potential to impact the lives of exponentially more.

Transparency Issues

Compounding the problem is the issue of transparency, or lack thereof. Much like Ambient Law, many of these algorithmic solutions are black-boxed—meaning that people reliant on their decisions have no way of knowing whether conclusions that have been reached are correct or the result of distorted or biased policy, or even of erroneous facts. Because of the step-by-step process at the heart of algorithms, codifying laws should make it more straightforward to examine audit trails about particular decisions, certainly when compared to dealing with a human. In theory, an algorithm can detail the specific rules that have been applied in each mini-decision, leading up to the final major one. In fact, the opacity of many automated systems means that they are shielded from scrutiny. For a variety of reasons, source code is not always released to the public. As a result, citizens are unable to see or debate new rules that are made; experiencing only the end results of decisions, as opposed to having access to the decision-making process itself. Individuals identified as potential terror suspects may receive questioning lasting many hours, or even be forced to miss flights, without ever finding out exactly why the automated system targeted them. This, in turn, means that there is a chance that they will be detained each time they attempt to board an airplane. In a Kafka-like situation, it is difficult to argue a certain conclusion if you do not know how it has been reached. It's one thing to have a formula theoretically capable of deciding particular laws, another entirely to have its inner workings transparent in a way that the general populace has access to it.

While these problems could, as noted, be the result of explicit bias, more often than not they are likely to be created accidentally by programmers with little in the way of legal training. As a result, there is a strong possibility that programmers might change the substance of particular laws when translating them into machine code. This was evidenced in a situation that occurred between September 2004 and April 2007, when programmers brought in from private companies embedded more than 900 incorrect rules within Colorado's public benefits system. "They got it so wrong that there were hundreds of thousands of incorrect assessments, because the policy embedded in the code was wrong," says University of Maryland law professor Danielle Citron. "It was all because they interpreted policy without a policy background."

The errors made by the coders included denying medical treatment to patients with breast and cervical cancer based upon income, as well as refusing aid to pregnant women. One

60-year-old, who had lost her apartment and was now living on the streets, found herself turned down for extra food stamps because she was not a "beggar," which is how coders had chosen to term "homelessness." In the end, eligibility workers for the Colorado Benefits Management System (CBMS) were forced to use fictitious data to get around the system's numerous errors.[60]

The Colorado situation went beyond erroneous computer code. The laws the programmers had entered were distorted to such a degree that they had effectively been changed. As Citron points out, were such amendments to have taken place within the framework of the legal system, they would have taken many months to push through the system. "If administrative law is going to pass a new policy regulation they may be required to put the policy up for comment, to have a notice period when they hear comments from interested advocates and policy-makers," she says. "They then incorporate those comments, provide an explicit explanation for their new policy, and respond to any comments made. Only then—after this extended notice and comment period, which is highly public in nature—can a new rule be passed. What happened here was a bypassing of democratic process designed to both allow participation and garner expertise from others." As it was, programmers were given vast and unreviewable policy-making power, defying any kind of judicial review.

While not all errors are as egregious as those denying treatment to cancer patients on the basis of income, a number of coded laws can lack the subtlety of their written counterparts: the result of either laziness or ignorance on the part of programmers. For instance, the United States' Food Stamp Act limits unemployed childless adults to three months of food stamps. However, it also provides six exceptions to this rule, which cross-reference other exceptions, which then refer to yet more exceptions. Since these exceptions are statistically rare, programmers may be tempted to write code that employs just the simplified three-month version of the rule, leaving out the complicated and potentially confusing exceptions.

While at present this is a problem that relates to the retrofitting of existing laws into code, longer term it presents other possibilities. It might be, for instance, that agencies will be increasingly inclined to adopt policies that favor simple questions and answers over more nuanced approaches—even when this would be the preferable option, since the former is easier to translate into algorithm than the latter. Going forward, this could result in complex laws being unnecessarily simplified so as to allow them to be better automated. Evidence of this is already being seen. In Massachusetts, IT specialists persuaded agency decision-makers to avoid adopting public benefits policies that would prove both challenging and expensive to automate. "Sometimes it feels like the IT department is running the policy," said one person close to the situation.[61]

Judge, Jury and Executable Code

Judge Richard Posner is the most cited legal scholar of the 20th century. Years ago, when Posner was a younger, less experienced judge, he presided over a patent case involving an early application of the kind of targeted recommendation technology Amazon would later carry out using algorithms. In the nascent days of satellite television, American television viewers experienced a seismic leap from having access to around 5 or 6

channels to up to 500. For many people, this was the birth of
the so-called paradox of choice. With so many options available, how could they possibly be expected to pick the channel
they most wanted to watch? One company came up with an
answer. Asking for a single channel in each home, they prom-
ised to send questionnaires to everyone who received the
channel, asking them to list the type of programs they watched
most regularly. On the basis of this questionnaire, the company
then claimed it would pick the programming they predicted
individual customers would most enjoy, thereby giving them
their own personalized television channels. "In the morning
they could work out that you might like to watch news,"
Posner recalls. "In the afternoon they might see that you like
to watch soap operas, and in the evening that you enjoy
watching horror movies." Why worry about the other 499
channels when there was one that would have everything you
wanted on it?

Posner thought the concept was "ingenious." The idea
stuck with him, and over the years he considered how it could
affect his own profession. "You could imagine a similar thing
being done with judges," he reflects. "You take their opinions,
their public statements, and whatever other information you
have about them and use it to create a profile that could function as a predictive device." Posner elaborated on the notion
in a recent paper entitled "The Role of the Judge in the
Twenty-First Century," in which he postulates on the ways in
which judicial practices will likely be altered by the arrival of
The Formula. "We are all familiar," he writes, "with how
Amazon.com creates and modifies reader profiles, and some
of us are familiar with data-mining, which is the same proced-

ure writ large—the computer identifies patterns and updates them as new data are received."

> I look forward to a time when computers will create profiles of judges' philosophies from their opinions and their public statements, and will update these profiles continuously as the judges issue additional opinions. [These] profiles will enable lawyers and judges to predict judicial behavior more accurately, and will assist judges in maintaining consistency with their previous decisions—when they want to.[62]

Judges can be prickly when it comes to the complexity of their decision-making process. "Our business is prophecy, and if prophecy were certain, there would not be much credit in prophesying," wrote Judge Max Radin in a 1925 essay entitled "The Theory of Judicial Decision: Or How Judges Think."[63] Since he was writing close to a century ago, Radin cannot entirely be blamed for underestimating the power of computation—and yet it is possible to take him to task for his judicial arrogance. After all, if a judge is objective in his thought processes and therefore not subject to hunches, biases or any other "processes . . . alien to good judges"[64] (as federal judge Joseph C. Hutcheson Jr. would note four years after Radin's essay was published), it stands to reason that their decision-making process should be apparent to anyone with the proper legal grounding.

As it turns out, even sufficient legal training may not be necessary, as was suggested in a 2004 study in which an algorithm competed with a team of legal experts to see which could predict the greater number of Supreme Court verdicts. The algorithm surprised the study's authors by correctly predicting 75 percent of the verdicts (based on only a handful of different metrics), as compared to the team of legal scholars, who guessed just 59 percent, despite having access to far more specialized information.[65] In its own way, the "Supreme Court Forecasting Project" was the legal profession's equivalent of IBM's Watson supercomputer winning $1 million on *Jeopardy!* in 2011— marking, as it did, the culmination of a long-held techno dream first proposed by the jurimetrics movement. In 1897, Oliver Wendell Holmes Jr. wrote enthusiastically of his belief that the legal system, as with science's natural laws, should be quantifiably predictable. "The object of our study . . . is prediction," he observed, "the prediction of the incidence of the public force through the instrumentality of the courts."[66]

But if anything, the Supreme Court Forecasting Project was a bastardization of the jurimetrics' utopian vision. Rather than demonstrating that the legal system's innate objectivity made it predictable, the 2004 study showed that the ability to correctly predict judges' verdicts resided in the fact that they were not objective, but rather filtered in each case through a combination of ideological preferences. For instance, one of the metrics the algorithm used to predict verdicts came down to whether judges voted Democrat or Republican, a factor that, objectively speaking, should no more influence whether a judge finds a defendant guilty or innocent than the length of time since they last ate.

It is these "attitudinal" biases that data-mining algorithms could be used to discover. "This could be a useful tool for judges, to help them be more self-aware when it comes to bias," Judge Posner says. A judge might, for instance, be soft on criminals, but tough on business fraud. "When they receive their profile they may become aware that they have certain unconscious biases that push them in certain directions," Posner continues.

Harry Surden, associate professor at the University of Colorado Law School and formerly a software engineer for Cisco Systems, agrees with Posner's proposal, but says that the revelations about the biases that affect judicial decision-making could shock the general public. "As a democratic society, we might not like what we see," he says.

One area where such algorithms would likely be embraced would be in the tool kit of lawyers, who could use them to accurately model a judge's behavior to shape witnesses and arguments for maximum impact in a courtroom. "This is probably going on today in some of the more sophisticated legal entities," Surden speculates. He points to the likes of hedge funds that make large bets whose outcomes are predicated on legal results. Using sophisticated legal modeling algorithms to predict verdicts in such a scenario could directly result in vast quantities of money being made.

Other legal scholars have also suggested that data mining could be used to reveal a granularity of wrongfulness in criminal trials, using machine-readable criteria to calculate the exact extent of a person's guilt. One imagines that it will not be too long before neuroscience (currently in the throes of an algorithmic turn) seeks to establish the exact degree to which a person is free to comply with the law, arguing over questions like determinism versus voluntarism. Data-mining tools may additionally have an application when it comes to meting out punishments: perhaps using algorithms to compare datasets of verdicts with datasets showing the postconviction behaviors of offenders. Could it be, for example, that there are precise lengths of ideal sentence to avoid wrongdoers relapsing into crime?

Pushing the idea yet further, what about the concept of automated judges? If we could construct a recommender system able to predict with 99 percent accuracy how a certain judge might rule on a particular case (perhaps even with greater levels of consistency than the human judge it is based on), would this help to make trials fairer? "In principle, it would be possible, although it's still a way away," Judge Posner says. "The main thing that would be left out would be how the judge's views changed with new information. Any change that may affect the way judges think would somehow have to be entered into the computer program, and weighed in order to decide how he would decide a case."

At least at present it takes the creativity of a (human) judge to resolve multiple parties' grievances, while also reconciling differing interpretations of the law. In this way, the judicial process is less about a mechanical objectivity than it is about a high level of intersubjective agreement: a seemingly minor, but crucial difference. Algorithms may have plenty of applications in the courtroom—and could even be used effectively to make the existing system fairer—but they're unlikely to start handing down sentences.

For now, at least.

CHAPTER 4

The Machine That Made Art

Quite possibly the most famous statement ever made about Hollywood is the one that screenwriter William Goldman laid out in his 1983 memoir, *Adventures in the Screen Trade*. Combing through his decades in the film industry for something approaching a profundity, all Goldman was able to muster was the idea that, when it comes to the tricky business of moviemaking, "Nobody knows anything. Did you know," he asks,

> that *Raiders of the Lost Ark* was offered to every single studio in town and they all turned it down? All except Paramount. Why did Paramount say yes? Because nobody knows anything. And why did all the other studios say no? Because nobody knows anything. And why did Universal, the mightiest studio of them all, pass on *Star Wars* . . . ? Because nobody . . . knows the least goddam [sic] thing about what is or isn't going to work at the box office.[1]

Goldman is hardly alone in his protestations. In the auto-biography of studio executive Mike Medavoy, the legendary film executive who had a hand in *Apocalypse Now*, *One Flew Over the Cuckoo's Nest* and *The Silence of the Lambs*, opines, "The movie business is probably the most irrational business in the world . . . [It] is governed by a set of rules that are abso-lutely irrational."[2] Hollywood lore is littered with stories of surefire winners that become flops, and surefire flops that become winners. There are niche films that appeal to everyone, and mainstream films that appeal to no one. Practically no Hol-lywood decision-maker has an unblemished record and, when one considers the facts, it's difficult to entirely blame them.

Take the ballad of the two A-list directors, for example. In the mid-2000s, James Cameron announced that he was busy working on a mysterious new script, called "Project 880." Cameron had previously directed a string of hit movies, including *The Terminator*, *Terminator 2*, *Aliens*, *The Abyss*, *True Lies* and *Titanic*, the latter of which won 11 Academy Awards and became the first film in history to earn more than a billion dollars. For the past ten years, however, he had been sidelined making documentaries. Moreover, his proposed new film didn't have any major stars, he was planning to shoot it in the experimental 3-D format, and he was asking for $237 mil-lion to do so. Perhaps against more rational judgment, the film was nonetheless made, and when it arrived in cinemas it was instantly labeled in the words of one reviewer, "The Most Expensive American Film Ever . . . And Possibly the Most Anti-American One Too."[3]

Would you have given Cameron the money to make it? The correct answer, as many cinema-goers will be aware, is yes. Retitled *Avatar*, Project 880 proceeded to smash the record set by *Titanic*, becoming the first film in history to earn more than $2 billion at the box office.

At around the same time that *Avatar* was gaining momentum, a second project was doing the rounds in Hollywood. This was another science-fiction film, also in 3-D, based upon a classic children's story, with a script cowritten by a Pulitzer Prize–winning author, and was to be made by Andrew Stanton, a director with an unimpeachable record who had previously helped create the highly successful Pixar films *WALL-E* and *Finding Nemo*, along with every entry in the acclaimed *Toy Story* series. Stanton's film (let's call it "Project X") came with a proposed $250 million asking price to bring to the screen—a shade more than Project 880. Project X received the go-ahead too, only this didn't turn out to be the next *Avatar*, but rather the first *John Carter*, a film that lost almost $200 million for its studio and resulted in the resignation of the head of Walt Disney Studios (the company who bankrolled it), despite the fact that he had only taken the job after the project was already in development. As per Goldman's Law, nobody knows anything.

Patterns Everywhere

This is, of course, exactly the type of conclusion that challenge-seeking technologists love to hear about. The idea that there should be something (the entertainment industry at that) that is entirely unpredictable is catnip to the formulaic mind. As it happens, blockbuster movies and high-tech start-ups do have a fair amount in common. Aside from the fact that most are flops, and investors are therefore reliant on the winners being sufficiently big that they more than offset the losers, there are few other industries where the power of the "elevator pitch" holds more sway. The elevator pitch is the idea that any popular concept should be sufficiently simple that it can be explained over the course of a single elevator ride. Perhaps not coincidentally, this is roughly the time frame of a 30-second commercial: the tool most commonly used for selling movies to a wide audience.

Like almost every popular idea, the elevator pitch (also known in Hollywood as the "high concept" phenomenon) has been attributed to a number of industry players, although the closest thing to an agreed-upon dictionary definition still belongs to Steven Spielberg. "If a person can tell me the idea in twenty-five words or less, it's going to make a pretty good movie," the director of *Jurassic Park* and *E.T. the Extra-Terrestrial* has said. "I like ideas . . . that you can hold in your hand."[4] Would you be in the least surprised to hear that Spielberg's father, Arnold Spielberg, was a pioneering computer scientist who designed and patented the first electronic library system that allowed the searching of what was then considered to be vast amounts of data? Steven Spielberg might have been a flop as a science student, but The Formula was in his blood.[5]

The Formula also runs through the veins of another Hollywood figure, whom *Time* magazine once praised for his "machine-like drive," and who originally planned to study engineering at MIT, before instead taking the acting route and going on to star in an almost unblemished run of box office smashes. Early in Will Smith's career, when he was little more than a fad pop star appearing in his first television show, *The*

Fresh Prince of Bel-Air, the aspiring thespian sat down with his manager and attempted to work out a formula that would transform him from a nobody into "the biggest movie star in the world." Smith describes himself as a "student of the patterns of the universe." At a time when he was struggling to be seen by a single casting director, Smith spent his days scrutinizing industry trade papers for trends that appeared in what global audiences wanted to see. "We looked at them and said, 'Okay, what are the patterns?'" he later recalled in an interview with *Time* magazine.[6] "We realized that *ten* out of ten had special effects. *Nine* out of ten had special effects with creatures. *Eight* out of ten had special effects with creatures and a love story . . ." Two decades later, and with his films having grossed in excess of $6.36 billion worldwide, Smith's methodology hasn't changed a great deal. "Every Monday morning, we sit down [and say], 'Okay, what happened this weekend, and what are the things that resemble things that have happened the last ten, twenty, thirty weekends?'" he noted.

Of course, as big a box-office attraction as Smith undoubtedly is, when it comes to universal pattern spotting, he is still strictly small-fry.

The Future of Movies

In the United Kingdom, there is a company called Epagogix that takes the Will Smith approach to movie prediction, only several orders of magnitude greater and without the movie star good looks. Operating out of a cramped office in Kennington, where a handful of data analysts sit hunched over their computers and the walls are covered with old film posters, Epagogix is the unlikely secret weapon employed by some of the biggest studios in Hollywood. Named after the Greek word for the path that leads from experience to knowledge, Epagogix carries a bold claim for itself: it can—so CEO and cofounder Nick Meaney claims—accurately forecast how much money a particular film is going to earn at the box office, before the film in question is even made.

Epagogix's route to Hollywood was something of an unusual one. Meaney, a fortysomething Brit with a mop of thick black hair and a face faintly reminiscent of a midcareer Orson Welles, had a background in risk management. During his career Meaney was introduced by several mathematician friends to what are called neural networks: vast, artificial brains used for analyzing the link between cause and effect in situations where this relationship is complex, unclear or both. A neural network could be used, for example, to read and analyze the sound recordings taken from the wheels of a train as it moves along railway tracks. Given the right information it could then predict when a particular stretch of track is in need of engineering, rather than waiting for a major crash to occur. Meaney saw that neural networks might be a valuable insurance tool, but the idea was quickly shot down by his bosses. The problem, he realized, is that insurance doesn't work like this: premiums reflect the actuarial likelihood that a particular event is going to occur. The better you become at stopping a particular problem from happening, the lower the premiums that can be charged to insure against it.

Movies, however, *could* benefit from that kind of out-of-the-box thinking. With the average production costs for a big budget movie running into the tens or even hundreds of millions of dollars, a suitably large flop can all but wipe out a movie studio.

This is exactly what happened in 1980, when the infamous turkey *Heaven's Gate* (singled out by *Guardian* critic Joe Queenan as the worst film ever made) came close to destroying United Artists. One studio boss told Meaney that if someone was able to come up with an algorithm capable of stopping a single money-losing film from being made each year, the overall effect on that studio's bottom line would be "immeasurable." Intrigued, Meaney set about bringing such a turkey-shooting formula to life. He teamed up with some data-analyst friends who had been working on an algorithm designed to predict the television ratings of hit shows. Between them they developed a system that could analyze a movie script on 30,073,680 unique scoring combinations—ranging from whether there are clearly defined villains, to the presence or lack of a sidekick character—before cranking out a projected box office figure.

To test the system, a major Hollywood studio sent Epagogix the scripts for nine completed films ready for release and asked them to use the neural network to generate forecasts for how much each would make. To complicate matters, Meaney and his colleagues weren't given any information about which stars the films would feature, who they were directed by, or even what the marketing budget was. In three of the nine cases, the algorithm missed by a considerable margin, but in the six others the predictions were eerily accurate. Perhaps the most impressive prediction of all concerned a $50 million film called *Lucky You*. *Lucky You* starred the famous actress Drew Barrymore, was directed by Curtis Hanson, the man who had made the hit movies *8 Mile* and *L.A. Confidential,* and was written by Eric Roth, who had previously penned the screenplay for *Forrest Gump.* It also concerned a popular subject: the world of

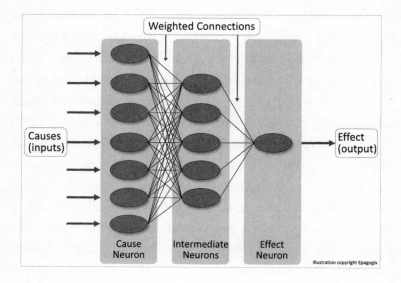

Illustration of a simplified neural network. The cat's cradle of connections in the middle, labeled the "intermediate neurons," is the proprietary "secret sauce" that makes Epagogix's system tick.

high-stakes professional poker. The studio was projecting big numbers for *Lucky You*. Epagogix's neural network, on the other hand, predicted a paltry $7 million. The film ended up earning just $6 million. From this point on, Epagogix began to get regular work.

A fair question, of course, is why it takes a computer to do this. As noted, Will Smith does something not entirely dissimilar from his kitchen table at the start of every week. Could a person not go through the 30,073,680 unique scoring combinations and check off how many of the ingredients a particular script adhered to? The simple answer to this is no.

While it would certainly be possible (albeit time-consuming) to plot each factor separately, this would say nothing about how the individual causal variables interact with one another to affect box office takings. To paraphrase a line from George Orwell's *Animal Farm*, all numbers might be equal, but some numbers are more equal than others.

Think about it in terms of a successful movie. In March 2012, *The Hunger Games* was released at cinemas and rapidly became a huge hit. But did *The Hunger Games* become a hit because it was based on a series of books, which had also been hits, and therefore had a built-in audience? Did it become a hit because it starred the actress Jennifer Lawrence, who *Rolling Stone* magazine once referred to as "the coolest chick in Hollywood"?[7] Or did it become a hit because it was released at a time of year when a lot of young people were out of school or college and were therefore free to go to the cinema? The best that anyone can say about any one of these questions is: maybe. *The Hunger Games* was based on a successful series of books, it did star a hot young actress popular with the film's key demographic, and it was released during the spring break holiday in the United States when large numbers of young people were on holiday. But the same could be said for plenty of other films that don't go on to become massive hits. And although *The Hunger Games* ultimately took more than $686 million in cinemas, how does anyone know whether all of the factors mentioned resulted in positive gains? Could it be that there was a potential audience out there who stayed home because they had heard that the film was based on a book, or because it starred Jennifer Lawrence, or because they knew that the cinema would be full of rowdy youths fresh out of school?

Might the film have earned a further $200 million if only its distributors had known to hold out until later in the year to release it?

These are the types of questions Epagogix seeks to quantify. A studio that employs Meaney and his colleagues will send Epagogix a shooting script, a proposed cast list, and a note about the specific time of year they plan to release their film. In return, they receive a sealed brown envelope containing the neural network's report. "We used to send reports that were this thick," Meaney says, creating a gap between his thumb and his forefinger to indicate a dossier the thickness of an average issue of *Vogue* magazine. Unconvinced that they were being read all the way through, the company now sends just two or three pages, bound by a single staple. "You might think that studios would want more than that, but in fact we spent a lot of time trimming these down," he continues.

The last page of the report is the most important one: the place where the projected box-office forecast for the film is listed. There is also a second, mysterious number: usually around 10 percent higher than the first figure, but sometimes up to twice its value. This figure is the predicted gross for the film on the condition that certain recommended tweaks are made to the script. Since regression testing is used to analyze each script in forensic detail, Meaney explains that the neural network can be used to single out individual elements where the potential yield is not where it should be—or where one part of the film is dragging down others. Depending upon your disposition and trust in technology, this is the point at which Epagogix takes a turn for either the miraculous or the unnerving. It is not difficult to imagine certain screenwriters would welcome the types of notes that might allow them to create a record-breaking movie, while others will detest the idea that an algorithm is telling them what to do.

It's not just scriptwriters who have the potential to be confused either. "One of the studio heads that we deal with on a regular basis is a very smart guy," Meaney says. "Early on in our dialogue with him he used to ask questions like, 'What would happen if the main character wears a red shirt? What difference does that make in your system?' He wasn't trying to catch us out; he was trying to grasp what we do. I was never able to explain to his satisfaction that it all depends. Is this a change that can be made without altering any of the other variables that the neural network ranks upon? Very seldom is there a movie where any significant alteration doesn't mean changes elsewhere." To modify a phrase coined by chaos theory pioneer Edward Lorenz, a butterfly that flaps its wings in the first minute of a movie may well cause a hurricane in the middle of the third act.

The studio boss to whom Meaney refers was likely picking a purposely arbitrary detail by mentioning the color of a character's shirt. After all, who ever formed their opinion about which movie to go and see on a Saturday night, or which film to recommend to friends, on the basis of whether the protagonist wears a blue shirt or a red shirt? But Meaney was nonetheless bothered by the comment: not because the studio boss was wrong, but because he wanted to reassure himself that saying "it depends" wasn't a cop-out. That evening he phoned one of his Epagogix colleagues back in the UK. "Quick as a flash, they said to me, 'Of course it depends.' Think about *Schindler's List*," Meaney recalls. "At the end of the film you get a glimpse of color and it's an absolutely pivotal moment in the narrative. In terms of our system it suddenly put the question into an historical context. In that particular film it makes the world of difference, while in another it might make no difference at all. Everything's relative."[8]

Parallel Universes

One way to examine whether there really are universal rules to be found in art would be to rewind time and see whether the same things became popular a second time around. If the deterministic formula that underpins companies like Epagogix is correct, then a blockbuster film or a best-selling novel would be successful no matter how many times you repeated the experiment over again. *Avatar* would always have grossed in excess of $2 billion, while *John Carter* would always have performed the exact same belly flop. Wolfgang Amadeus Mozart was always destined for greatness, while Antonio Salieri was always doomed to be an also-ran. This would similarly mean that there is no such thing as an "unlikely hit," since universal truths would state that there are rules that define something as a success or a failure. Adhere to them and you have a hit. Fail to do so and you have a flop.

A few years ago, a group of researchers from Princeton University came up with an ingenious way of testing this theory using computer simulations. What sparked Matthew Salganik, Peter Dodds and Duncan Watts's imagination was the way in which they saw success manifest itself in the entertainment industry. Much as some companies start out in a crowded field and go on to monopolize it, so too did they notice that par-

THE MACHINE THAT MADE ART 173

ticular books or films become disproportionate winners. These are what are known as "superstar" markets. In 1997, for instance, *Titanic* earned nearly 50 times the average U.S. box-office take for a film released that year. Was *Titanic* really 50 times better than any other film released that year, the trio wondered, or does success depend on more than just the intrinsic qualities of a particular piece of content? "There is tremendous unpredictability of success," Salganik says. "You would think that superstar hits that go on to become so successful would be somehow different from all of the other things they're competing against. But yet the people whose job it is to find them are unable to do so on a regular basis."

To put it another way, if Goldman's Law that nobody knows anything is right, is this because experts are too stupid to realize a hit when they have one on their hands, or does nobody know anything because nothing is for certain?

In order to test their hypothesis, Salganik, Dodds and Watts created what they referred to as an "Experimental Study of Inequality and Unpredictability in an Artificial Cultural Market."[9] This consisted of an online music market, a bit like iTunes, but featuring unknown songs from unknown bands. The 14,341 participants recruited online were given the chance to listen to the songs and rate them between 1 ("I hate it") and 5 ("I love it"). They could then download those songs that they liked the most. The most downloaded tracks were listed in a Top 40–style "leader board" that was displayed in a prominent position on the website.

What made this website different from iTunes was that Salganik, Dodds and Watts had not created one online music market, but many. When users logged on to the site, they were randomly redirected to one of nine "parallel universes" that were identical in every way with the exception of the leader board. According to the researchers' logic, if superstar hits really were orders of magnitude better than average, one would expect the same songs to appear in the same spot in each universe.

What Salganik, Dodds and Watts discovered instead was exactly what they had suspected: that there is an accidental quality to success, in which high-ranking songs take an early lead for reasons that seem inconsequential, based upon those taste-makers who sample it first. Once this lead is established, it is exacerbated through social feedback. A bookshop, for instance, might notice that one particular book is proving more popular than others, and therefore decide to order more copies. At this stage, the book may be selling 11 copies for every 10 sold by its next most popular rival—a marginal improvement. But when the new copies arrive they are displayed in favorable places around the shop (on a table next to the front door, for example) and soon the book is selling twice as many copies as its closest rival. To sell even more, the bookshop then decides to try to attract new customers by lowering its own profit margins and selling the book at a reduced price. At this point the book is selling four times as many copies as its closest rival. Because customers have the impression that the book is popular (and therefore must be good) they are more likely to buy it, thereby driving sales up even more.

This is what psychologists call the "mere-exposure" effect. At a certain juncture a tipping point is reached, where people will buy copies of the book so as not to be left out of what they see as a growing phenomenon, in much the same way that we might tune in to an episode of a television show that has gained a lot of buzz, just to see what all the fuss is about.

In Salganik, Dodds and Watts's experiment the songs that were ranked as the least popular in one universe would never prove the most popular in another, while the most popular songs in one universe would never prove the least popular somewhere else. Beyond this, however, any other result was possible.

The Role of Appeal

As you may have detected, there was a problem posed by the formulation of Salganik, Dodds and Watts, one that they acknowledged when it came time to write up their findings. In an experiment designed to determine the relationship between popularity and quality, how could any meaningful conclusions be drawn without first deciding upon a quantifiable definition for quality? "Unfortunately," as the three researchers gravely noted in their paper, "no generally agreed upon measure of quality exists, in large part because quality is largely, if not completely, a social construction." To get around this issue (which they referred to as "these conceptual difficulties") Salganik, Dodds and Watts chose to eschew debate about the artistic "quality" of individual songs altogether, and instead to focus on the more easily measurable characteristic of "appeal."

A song's appeal was established through the creation of one more music market, this time with no scoreboard visible. Lacking the presence of any obvious social feedback mechanisms, Salganik, Dodds and Watts theorized that whichever song turned out to be the most popular in this scenario would do so based wholly on the merits of its objective appeal. What those were didn't matter. All that mattered was that they existed.

This market-driven reading of "appeal" over "quality" is (no pun intended) a popular one. "There are plenty of films that, to me, might be better than *Titanic*, but in the marketplace it's *Titanic* that earns the most," says Epagogix's Nick Meaney. If what emanates from his company's neural network happens to coincide with what Meaney considers a great work of art, that is wonderful. If it doesn't, it's better business sense to recommend studios fund a film that a lot of people will pay money to see and not feel cheated by, rather than one that a few critics might rave about but nobody else will watch. Netflix followed a similar logic to Meaney when in 2006 it implemented its (now abandoned) $1,000,000 open competition to ask users to create a filtering algorithm that markedly improved upon Netflix's own recommender system. Instead of an "improvement" being an algorithm that directed users toward relatively obscure critical favorites like Yasujirô Ozu's 1953 masterpiece *Tokyo Story* or Jean Renoir's 1953 *La Règle du jeu*, Netflix judged "better recommendations" as recommendations that most accurately predicted the score users might give to a film or TV show. To put it another way, this is "best" in the manner of the old adage stating that the best roadside restaurants are those with the most cars parked outside them.

The advent of mass appeal is a fairly modern concept, belonging to the rise of factory production lines in the late 19th and early 20th century. For the first time in history, a true mass market emerged as widespread literacy coincided with the large-scale move of individuals to cities. This was the birth of the packaged formula, requiring the creation of products designed to sell to the largest number of people possible. Mass production also meant standardization, since the production and distribution process demanded that everything be reduced to its simplest possible components. This mantra didn't just apply to consumer goods, but also to things that didn't inherently require simplification as part of their production process. As the authors of newspaper history *The Popular Press, 1833–1865* observe, for example, the early newspaper tycoons "packaged news as a *product* [my emphasis] to appeal to a mass audience."[10]

In this way, art, literature and entertainment were no different from any other product. In the dream of the utopianists of the age, mass appeal meant that the world would move away from the elitist concept of "art" toward its formalized big brother, "engineering." Cars, airplanes and even entire houses would roll off the factory conveyor belt en masse, signaling an end to an existence in which inequality was commonplace. How could it, when everyone drove the same Model T Ford and lived in the same homes? Art was elitist, irrational and superficial; engineering was collectivist, functional and hyper-rational. Better to serve the democratized objectivity of the masses than the snobbish subjectivity of the few.[11]

It was cinema that was seized upon as the ideal medium for conveying popular, formulaic storytelling, representing the first example of what computational scholar Lev Manovich refers to as New Media. "Irregularity, nonuniformity, the accident and other traces of the human body, which previously, inevitably accompanied moving image exhibitions, were replaced by the uniformity of machine vision," Manovich writes.[12] In cinema, its pioneers imagined a medium that could apply the engineering formula to the field of entertainment. Writing excitedly about the bold new form, Soviet filmmaker and propagandist Sergei Eisenstein opined, "What we need is science, not art. The word *creation* is useless. It should be replaced by *labor*. One does not create a work, one constructs it with finished parts, like a machine."

Eisenstein was far from alone in expressing the idea that art could be made more scientific. A great many artists of the time were similarly inspired by the notion that stripping art down to its granular components could provide their work with a social function on a previously unimaginable scale, thereby achieving the task of making art "useful." A large number turned to mechanical forms of creativity such as textile, industrial and graphic design, along with typography, photography and photomontage. In a state of euphoria, the Soviet artists of the Institute for Artistic Culture declared that "the last picture has been painted" and "the 'sanctity' of a work of art as a single entity . . . destroyed." Art scholar Nikolai Punin went one step further still, both calling for and helpfully creating a mathematical formula he claimed to be capable of explaining the creative process in full.[13]

Unsurprisingly, this mode of techno-mania did not go unchallenged. Reacting to the disruptive arrival of the new technologies, several traditionally minded scholars turned their attentions to critiquing what they saw as a seismic shift in the world of culture. For instance, in his essay "The Work of Art in the Age of Mechanical Reproduction," German philosopher and literary critic Walter Benjamin observed:

With the advent of the first truly revolutionary means of reproduction, photography, . . . art sensed the approaching crisis . . . Art reacted with the doctrine of *l'art pour l'art*, that is, with a theology of art. This gave rise to . . . "pure" art, which not only denied any social function of art but also any categorizing by subject matter.[14]

Less than a decade later in 1944, two German theorists named Theodor Adorno and Max Horkheimer elaborated on Benjamin's argument in their *Dialectic of Enlightenment*, in which they attacked what they bitingly termed the newly created "culture industry." Adorno and Horkheimer's accusation was simple: that like every other aspect of life, creativity had been taken over by industrialists obsessed with measurement and quantification. In order to work, artists had to conform, kowtowing to a system that "crushes insubordination and makes them subserve the formula."

Had they been alive today, Adorno and Horkheimer wouldn't for a moment have doubted that a company like Epagogix could successfully predict the box office of Hollywood movies ahead of production. Forget about specialist reviewers; films planned from a statistical perspective call for nothing more or less than statistical analysis.

Universal Media Machines

In 2012, another major shift was taking place in the culture industry. Although it was barely remarked upon at the time, this was the first year in which U.S. viewers watched more films legally delivered via the Internet than they did using physical formats such as Blu-ray discs and DVDs. Amazon, meanwhile, announced that, less than two years after it first introduced the Kindle, customers were now buying more e-books than they were hardcovers and paperbacks combined.

At first glance, this doesn't sound like such a drastic alteration. After all, it's not as if customers stopped watching films or reading books altogether, just that they changed something about the way that they purchased and consumed them. An analog might be people continuing to shop at Gap, but switching from buying "boot fit" to "skinny" jeans.

However, while this analogy works on the surface, it fails to appreciate the extent of the transition that had taken place. It is not enough to simply say that a Kindle represents a book read on screen as opposed to on paper. Each is its own entity, with its own techniques and materials. In order to appear on our computer screens, tablets and smartphones, films, music, books and paintings must first be rendered in the form of digital code. This can be carried out regardless of whether a particular work was originally created using a computer or not. For the first time in history, any artwork can be described in mathematical terms (literally a formula), thereby making it programmable and subject to manipulation by algorithms.

In the same way that energy can be transferred from movement into heat, so too can information now shift easily between mediums. For instance, an algorithm could be used to identify the presence of shadows in a two-dimensional photograph and then translate these shadows to pixel depth by measuring their position on the grayscale—ultimately outputting a three-dimensional object using a 3-D printer.[15] Even more impressively, in recent years Disney's R&D division has been hard at work on a research project designed to simulate the feeling of touching a real, tactile object when, in reality, users are only touching a flat touch screen. This effect is achieved using a haptic feedback algorithm that "tricks" the brain into thinking it is feeling ridges, bumps or potentially even textures, by re-creating the sensation of friction between a surface and a fingertip.[16] "If we can artificially stretch skin on a finger as it slides on the touch screen, the brain will be fooled into thinking an actual physical bump is on a touch screen even though the touch surface is completely smooth," says Ivan Poupyrev, the director of Disney Research, who describes the technology as a means by which interactions with virtual objects can be made more realistic.[17]

There is also the possibility of combining different mediums in entirely new ways, something increasingly common in a world used to web pages, PowerPoint presentations, and mobile multimedia messages. It is no coincidence that the advent of the programmable computer in the 20th century saw the art world take its first tentative steps away from the concept of media specificity. As the computer became a multipurpose canvas for everything from illustration to composition, so too did modern artists over the past 50 years seek to establish formulas capable of bringing together previously separate entities, such as musical and visual composition.

Scientists and artists alike have long been fascinated by the neurological condition of synesthesia (Greek for "joined perception"), in which affected individuals see words as colors, hear sounds as textures, or register smells as shapes. A similar response is now reproducible on computer, and this can be seen through the increasing popularity of "info-aesthetics"[18] that has mirrored the rise of data analytics. More than just a method of computing, info-aesthetics takes numbers, text, networks, sounds and video as its source materials and re-creates them as images to reveal hidden patterns and relationships in the data.

Past data visualizations by artists include the musical compositions of Bach presented as wave formations, the thought processes of a computer as it plays a game of chess, and the fluctuations of the stock market. In 2013, Bill Gates and Microsoft chief technology officer Nathan Myhrvold filed a patent for a system capable of taking selected blocks of text and using this information to generate still images and even full-motion video. As they point out, such technology could be of use in a classroom setting—especially for students suffering from dyslexia, attention deficit disorder, or any one of a number of other conditions that might make it difficult to read long passages of text.[19]

To Thine Own Self Be True/False

Several years ago, as an English graduate student, Stephen Ramsay became interested in what is known as graph theory. Graph theory uses the mathematical relationship between objects to model their connections—with individual objects represented by "nodes" and the lines that connect them referred to as "edges." Looking around for something in literature that was mathematical in structure, Ramsay settled upon the plays of William Shakespeare. "A Shakespearean play will start in one place, then move to a second place, then go back to the first place, then on to the third and fourth place, then back to the second, and so on," he says. Intrigued, Ramsay set about writing a computer program capable of transforming any Shakespearean play into a graph. He then used data-mining algorithms to analyze the graphs to see whether he could predict (based wholly on their mathematical structure) what he was looking at was a comedy, tragedy, history or romance. "And here's the thing," he says. "I could. The computer knew that *The Winter's Tale* was a romance, it knew that *Hamlet* was a tragedy, it knew that *A Midsummer Night's Dream* was a comedy." There were just two cases in which the algorithm, in Ramsay's words, "screwed up." Both *Othello* and *Romeo and Juliet* came back classified as comedies. "But this was the part that was flat-out amazing," he says. "For a number of years now, literary critics have been starting to notice that both plays have the structure of comedies. When I saw the conclusion the computer had reached, I almost fell off my chair in amazement."

The idea that we might practically use algorithms to find the "truths" obscured within particular artistic works is not a new one. In the late 1940s, an Italian Jesuit priest named Roberto Busa used a computer to "codify" the works of influential theologian Thomas Aquinas. "The reader should not simply attach to the words he reads the significance they have in his mind," Busa explained, "but should try to find out what significance they had in the author's mind."[20]

Despite this early isolated example, however, the scientific community of the first half of the 20th century for the most part doubted that computers had anything useful to say about something as unquantifiable as art. An algorithm could never, for example, determine authorship in the case of two painters with similar styles—particularly not in situations in which genuine experts had experienced difficulty doing so. In his classic book *Faster Than Thought: A Symposium on Digital Computing Machines*, the late English scientist B. V. Bowden offers the view that:

> It seems most improbable that a machine will ever be able to give an answer to a general question of the type: "Is this picture likely to have been painted by Vermeer, or could van Meegeren have done it?" It will be recalled that this question was answered confidently (though incorrectly) by the art critics over a period of several years.[21]

To Bowden, the evidence is clear, straightforward and damning. If Alan Turing suggested that the benchmark of an intelligent computer would be one capable of replicating the intelligent actions of a man, what hope would a machine have of resolving a problem that even man was unable to make an intelligent judgment on? A cooling fan's chance in hell, surely.

In recent years, however, this view has been challenged. Lior Shamir is a computer scientist who started his career working for the National Institutes of Health, where he used robotic microscopes to analyze the structure of hundreds of thousands of cells at a time. After that he moved on to astronomy, where he created algorithms designed for scouring images of billions of galaxies. Next he began working on his biggest challenge to date: creating the world's first fully automated, algorithmic art critic, with a rapidly expanding knowledge base and a range of extremely well-researched opinions about what does and does not constitute art. Analyzing each painting it is shown based on 4,024 different numerical image content descriptors, Shamir's algorithm studies everything that a human art critic would examine (an artist's use of color, or their distribution of geometric shapes), as well as everything that they probably wouldn't (such as a painting's description in terms of its Zernike polynomials, Haralick textures and Chebyshev statistics). "The algorithm finds patterns in the numbers that are typical to a certain artist," Shamir explains.[22] Already it has proven adept at spotting forgeries, able to distinguish between genuine and fake Jackson Pollock drip paintings with an astonishing 93 percent accuracy.

Much like Stephen Ramsay's Shakespearean data-mining algorithm, Shamir's automated art critic has also made some fresh insights into the connections that exist between the work of certain artists. "Once you can represent an artist's work in terms of numbers, you can also visualize the distance between their work and that of other artists," he says. When analyzing the work of Pollock and Vincent Van Gogh—two artists who worked within completely different art movements—Shamir discovered that 19 of the algorithm's 20 most informative descriptors showed significant similarities, including a shared preference for low-level textures and shapes, along with a similar deployment of lines and edges.[23] Again, this might appear to be a meaningless insight were it not for the fact that several influential art critics have recently begun to theorize similar ideas.[24]

Bring on the Reading Machines

This newfound ability to subject media to algorithmic manipulation has led a number of scholars to call for a so-called algorithmic criticism. It is no secret that the field of literary studies is in trouble. After decades of downward trends in terms of enrollments, the subject has become a less and less significant part of higher education. So how could this trend be reversed? According to some, the answer is a straightforward one: by turning it into the "digital humanities," of course. In a 2008 editorial for the *Boston Globe* entitled "Measure for Measure," literary critic Jonathan Gottschall dismissed the current state of his field as "moribund, aimless, and increasingly irrelevant to the concerns . . . of the 'outside world.'" Forget about vague terms like the "beauty myth" or Roland Barthes's concept of the death of the author, Gottschall says. What is needed instead is a productivist approach to media built around correlations, pattern-seeking and objectivity.

As such, Gottschall lays out his Roberto Busa–like beliefs that genuine, verifiable truths both exist in literature and are desirable. In keeping with the discoverable laws of the natural sciences, in Gottschall's mind there are clear right and wrong answers to a question such as, "Can I interpret [this painting/ this book/this film] in such-and-such a way?"

While these comments are likely to shock many of those working within the humanities, Gottschall is not altogether wrong in suggesting that there are elements of computer science that can be usefully integrated into arts criticism. In the world of The Formula, what it is that is possible to know changes dramatically. For example, algorithms can be used to determine "vocabulary richness" in literature by measuring the number of different words that appear in a 50,000-word block of text. This can bring about a number of surprises. Few critics would ever have suspected that a "popular" author like Sinclair Lewis—sometimes derided for his supposed lack of style—regularly demonstrates twice the vocabulary of Nobel laureate William Faulkner, whose work is considered notoriously difficult.

One of the highest-profile uses of algorithms to analyze text took place in 2013 when a new crime fiction novel, *The Cuckoo's Calling*, appeared on bookshelves around the world, written by a first-time author called Robert Galbraith. While the book attracted little attention early on, selling just 1,500 printed copies, it became the center of controversy after a British newspaper broke the story that the author may be none other than *Harry Potter* author J. K. Rowling, writing under a pseudonym. To prove this one way or the other, computer scientists were brought in to verify authorship. By using data-mining techniques to analyze the text on four different variables (average word length, usage of common words, recurrent word pairings, and distribution of "character 4-grams"), algorithms concluded that Rowling was most likely the author of the novel, something she later admitted to.[25]

As Stephen Ramsay observes, "The rigid calculus of computation, which knows nothing about the nature of what it's examining, can shock us out of our preconceived notions on a particular subject. When we read, we do so with all kinds of biases. Algorithms have none of those. Because of that they can take us off our rails and make us say, 'Aha! I'd never noticed that before.'"

Data-tainment

A quick scan of the best-seller list will be enough to convince us that, for better or worse, book publishers are not the same as literary professors. This doesn't mean that they are exempt from the allure of using algorithms for analysis, however. Publishers, of course, are less interested in understanding a particular text than they are in understanding their customers. In previous years, the moment that a customer left a bookshop and took a book home with them, there was no quantifiable way a publisher would know whether they read it straight through or put it on a reading pile and promptly forgot about it. Much the same was true of VHS tapes and DVDs. It didn't matter how many times an owner of *Star Wars* rewound their copy of the tape to watch a Stormtrooper bump his head, or paused *Basic Instinct* during the infamous leg-crossing scene: no studio executive was ever going to know about it. All of that is now changing, however, due to the amount of data that is able to be gathered and fed back to content publishers. For example, Amazon is able to tell how quickly its customers read e-books, whether they scrutinize every word of an introduction or skip over it altogether, and even which sections they choose to highlight. They know that science fiction, romance and crime novels tend to be read faster than literary fiction, while nonfiction books are less likely to be finished than fiction ones.

These insights can then be used to make creative decisions. In February 2013, Netflix premiered *House of Cards*, its political drama series starring Kevin Spacey. On the surface, the most notable aspect of *House of Cards* appeared to be that Netflix—an on-demand streaming-media company—was changing its business model from distribution to production, in an effort to compete with premium television brands like Showtime and HBO. Generating original video content for Internet users is still something of a novel concept, particularly when it is done on a high budget and, at $100 million, *House of Cards* was absolutely that. What surprised many people, however, was how bold Netflix was in its decisions. Executives at the Los Gatos–based company commissioned a full two seasons, comprising 26 episodes in total, without ever viewing a single scene. Why? The reason was that Netflix had used its algorithms to comb through the data gathered from its 25 million users to discover the trends and correlations in what people watched. What it discovered was that a large number of subscribers enjoyed the BBC's *House of Cards* series, evidenced by the fact that they watched episodes multiple times and in rapid succession. Those same users tended to also like films that starred Kevin Spacey, as well as those that were directed by *The Social Network*'s David Fincher. Netflix rightly figured that a series with all three would therefore have a high probability of succeeding.[26]

The gamble appeared to pay off. Under a review titled "*House of Cards* Is All Aces," *USA Today* praised the show as "money well-spent" and among the "most gorgeous [pieces] of television" people were likely to see all year.[27] President Obama admitted to being a fan. Netflix followed up its *House of Cards* success with three more well-received series: *Hemlock Grove*, *Arrested Development* and *Orange Is the New Black*. At the 2013 Emmy Awards, the company notched up a total of 14 nominations for its efforts.[28] "It took HBO 25 years to get its first Emmy nomination," noted American TV critic and columnist David Bianculli in an article for the *New York Times*. "It took Netflix six months."[29]

Netflix's success has seen it followed by online retailer Amazon, which also has access to a vast bank of customer information, revealing the kind of detailed "likes" and "dislikes" data that traditional studio bosses could only dream of. "It's a completely new way of making movies," Amazon founder Jeff Bezos told *Wired* magazine. "Some would say our approach is unworkable—we disagree."[30]

In Soviet Russia, Films Watch You

In a previous life, Alexis Kirke worked as a quantitative analyst on Wall Street, one of the so-called rocket scientists whose job concerns a heady blend of mathematics, high finance and computer skills. Having completed a PhD in computer science, Kirke should have been on top of the world. "Quants" are highly in demand and can earn upward of $250,000 per year, but Kirke nonetheless found himself feeling surprisingly disenfranchised. "After about a year, I decided that this wasn't what I wanted to do," he says. What he wanted instead was to pursue an artistic career. Kirke left the United States, moved back home to Plymouth, England, and enrolled in a music degree course. Today, he is a research fellow at Plymouth University's Interdisciplinary Center for Computer Music Research.

In 2013, Kirke achieved his greatest success to date when he created *Many Worlds*, a film that changed the direction of its narrative based upon the response of audience members. *Many Worlds* premiered at the Peninsula Arts Contemporary Music Festival in 2013, and its interactivity marked a major break from traditional cinema by transforming audiences from passive consumers into active participants. At screenings, audience members were fitted with special sensors capable of monitoring their brain waves, heart rate, perspiration levels and muscle tension. These indicators of physical arousal were then fed into a computer, where they were averaged and analyzed in real time, with the reactions used to trigger different scenes. A calm audience could conceivably be jolted to attention with a more dramatic sequence, while an already tense or nervous audience could be shown a calmer one. This branching narrative ultimately culminated in one of four different endings.[31]

In a sense, companies like Epagogix, which I mentioned at the start of the chapter, offer a new twist on an old idea: that there is such a thing as a work of art that will appeal to everyone. Anyone who has ever read two opposite reviews of the same film—one raving about it and the other panning it—will realize that this is not necessarily true. Our own preferences are based on synthetic concepts based around inherited ideas, as well as our own previous experiences. My idea of how *Macbeth* should be performed on the stage is based on those performances I have attended in the past, or what I have read about the play. The same is true of the films I like, the music I enjoy, and the books I read.

"A fixed film appeals to the lowest common denominator," says Kirke. "What it does is to plot an average path through the audience's emotional experience, and this has to translate across all audiences in all countries. Too often this can end in compromise." Kirke isn't wrong. For every *Iron Man*—a Hollywood blockbuster that appeals to vast numbers without sacrificing quality—there are dozens of other films from which every ounce of originality has been airbrushed in an effort to appease the widest possible audience. *Many Worlds* suggests an alternative: that in the digital age, rationalization no longer has to be the same as standardization. Formulas can exist, but these don't have to ensure that everything looks the same.[32]

A valid question, of course, concerns the cost of implementing this on a wider level. Alexis Kirke created *Many Worlds* on what he describes as a "nano-budget" of less than $4,000—along with some lights, tripods and an HD camera borrowed from Plymouth University's media department for a few days. How would this work when scaled up to Hollywood levels? After all, at a time when blockbuster movies can cost upward of $200 million, can studios really afford the extra expenditure of shooting four different endings in the way that Kirke did? He certainly believes they can. As I described earlier in this chapter, the entertainment industry currently operates on a highly inefficient (some would say unscientific) business model reliant on statistically rare "superstar" hits to offset the cost of more likely losses. The movie studio that makes ten films and has two of these become hits will be reasonably content. But what if that same studio ramped up its spending by shooting alternate scenes and commissioning several possible sound tracks at an additional cost of 50 percent per film, although this in turn meant that the film was more likely to become a hit? If branching films could be all things to all people, studios might only have to make five films to create two sizeable hits.

Following the debut of *Many Worlds*, Kirke was approached by several major media companies interested in bringing him on board to work as a consultant. The BBC twice invited him to its headquarters in Manchester to screen the film and discuss his thoughts on the future of interactive media. Manufacturers were particularly interested in how this technology could usefully be integrated into the next generation of television sets. "This is something that's already starting to happen," Kirke says. In 2013, Microsoft was awarded a patent for a camera cap-

able of monitoring the behavior of viewers, including movement, eye tracking and heart rate. This behavior can then be
compiled into user-specific reports and sent, via the cloud, to a
remote device able to determine whether certain goals have
been met.[33] Advertisers, for instance, will have the option of
rewarding viewers who sit through commercial breaks with
digital credits (iTunes vouchers, perhaps) or physical prizes.
Because Microsoft's camera sensor has the ability to recognize
gestures, advertisers could create dances or actions for viewers
to reproduce at home. The more enthusiastic the reproduction,
the more iTunes vouchers the viewer could win.

Another company, named Affectiva, is beginning to market
facial expression analysis software to the consumer product manufacturers, retailers, marketers and movie studios. Its mission is
to mine the emotional response of consumers to help improve
the designs and marketing campaigns of products.[34] Film and
television audiences will similarly increasingly be watched by
nonspeech microphones and eye line sensors, along with social
network scanners built into mobile devices, which adjust whatever they are watching according to reactions. If it is determined
that a person's eyes are straying from the screen too often, or
that they are showing more interest in Facebook than the entertainment placed in front of them, films will have the option of
adjusting editing, sound track or even narrative to ensure that
maximum engagement level is maintained at all times.

A Moving Target

Traditionally, the moment that a painting was finished, a
photograph was printed or a book was published it was fixed

in place. We might even argue that such a quality forms part of our appreciation. With its fixed number of pages bound by a single spine, the physical organization of a book invites the reader to progress through it in a linear, predetermined manner—moving from left to right across the page, then from page to page, and ultimately from chapter to chapter, and cover to cover.[35] As a result, a book appeals to our desire for completion, wholeness and closure.

No such permanence or fixedness exists in the world of The Formula, in which electronic books, films and music albums can be skipped through at will.[36] This, in turn, represents a flattening of narrative, or a division of it into its most granular elements. As computer scientist Steven DeRose argues in a 1995 paper entitled "Structured Information: Navigation, Access and Control," this analysis of structured information does not get us close to certain universal truths, "in the sense that a Sherlock Holmes should peer at it and discern hidden truth . . . but rather in the sense that the information is divided into component parts, which in turn have components, and so on."[37]

This narrative unwinding was demonstrated to great effect several years ago when the American artist Jason Salavon digitized the hit movie *Titanic* and broke it up into its separate frames. Each of these frames was then algorithmically averaged to a single color using a computer, before the frames were recollected as a unified image, mirroring the narrative sequence of the film. Reading the artwork from left to right and top to bottom, the movie's rhythm was laid out in pure color.[38]

Both Alexis Kirke's *Many Worlds* and Salavon's reimagining of *Titanic* represent two sides of the same coin. In a post-9/11 age in which our own sense of impermanence is heightened, past and present are flattened in the manner of a Facebook timeline, and the future is an uncertain prospect, what relevance do traditional beginnings, middles and ends have? This is further seen by the number of artworks that, imbued with the power of code and real-time data streams, exist in a state of constant flux. In the same way that the Internet will never be completed—any more than technology itself can be completed—these algorithmic artworks are able to adapt and mutate as new data inputs are absorbed into the whole.

An example of this was created in Cambridge, Massachusetts, where two members of Google's Big Picture data visualization group, Fernanda Viégas and Martin Wattenberg, coded an online wind map of the United States, which presents data from the National Digital Forecast Database in the hypnotic form of a swirling, constantly changing animation.[39] "On calm days it can be a soothing meditation on the environment," Wattenberg says. "During hurricanes it can become ominous and frightening."[40] In a previous age of fixedness, a work of art became timeless by containing themes universal enough to span generations. Today "timeless" means changing for each successive audience: a realization of the artist's dilemma that work is never finished, only abandoned.

In his latest book, *Present Shock*, cyberpunk media theorist Douglas Rushkoff seizes upon a similar idea to discuss the ways in which today's popular culture reflects The Formula. Much as the artists of the early 20th century adopted the techniques and aesthetics of heavy-duty industrial machinery as their model of choice for the direction in which to take their art, so too does today's entertainment industry reflect the flux-like possibilities of code. Unlike the predictable narrative character arcs of classic films like *The Godfather*, today's most lauded creations are ongoing shows such as *Game of Thrones* that avoid straightforward, three-act structures and simply continue indefinitely.

Looking at shows like the NBC series *Community* and Seth MacFarlane's *Family Guy*, Rushkoff further demonstrates the technology-induced collapse of narrative structure at work. *Community* features a group of misfits at Greendale Community College, who constantly refer to the fact that they are characters within a sitcom. What story arcs do exist in the show are executed with the full knowledge that the viewing audience is well versed in the clichés that make up most traditional sitcoms. *Family Guy* similarly breaks away from traditional narrative storytelling in favor of self-contained "cutaway" gags, which prove equally amusing regardless of the order in which they are played, making it the perfect comedy for the iPod Shuffle generation. Like its obvious forerunner, *The Simpsons*, rarely does a plot point in *Family Guy* have any lasting impact—thereby allowing all manner of nonsensical occurrences to take place before the "reset" button is hit at the end of each episode.

A more poignant illustration of this conceit can be found in the more serious drama series on television. Shows like *The Wire*, *Mad Men*, *The Sopranos* and *Dexter* all follow ostensibly different central characters (ranging from Baltimore police and Madison Avenue admen to New Jersey mobsters and Miami serial killers) whose chief similarity is their inability to change their nature, or the world they inhabit. As Rushkoff writes, these series don't work their magic through a linear plot, but instead create contrasts through association, by nesting screens within screens, and by giving viewers the tools to make connections between various forms of media . . . The beginning, the middle, and the end have almost no meaning. The gist is experienced in each moment as new connections are made and false stories are exposed or reframed. In short, these sorts of shows teach pattern recognition, and they do it in real time.[41]

Even today's most popular films no longer exist as unitary entities, but as nodes in larger franchises—with sequels regularly announced even before the first film is shown. It's no accident that in this setting many of the most popular blockbusters are based on comic-book properties: a medium in which, unlike a novel, plot points are ongoing with little expectation of an ultimate resolution.

In this vein, Alexis Kirke's *Many Worlds* does not exist as an experimental outlier, but as another step in the unwinding of traditional narrative and a sign of things to come. While stories aren't going anywhere, Kirke says, in the future audiences are likely to be less concerned with narrative arcs than they will with emotional ones.

Digital Gatekeepers

A lack of fixedness in art and the humanities can have other, potentially sinister, implications. Because the "master" copy of a particular book that we are reading—whether this be on Kindle or Google Books—is stored online and accessed via "the cloud," publishers and authors now possess the ability to make changes to works even after they have been purchased and taken home. A poignant illustration of this fact occurred in 2009 when Amazon realized that copies of George Orwell's classic novel *Nineteen Eighty-Four* being sold through its Kindle platform were under copyright, rather than existing in the public domain as had been assumed. In a panic, Amazon made the decision to delete the book altogether, resulting in it vanishing from the libraries of all those who had purchased it. The irony, of course, is that *Nineteen Eighty-Four* concerns a dystopian future in which the ruling superpower manipulates its populace by rewriting the history books on a daily basis. More than 60 years after the novel was first published, such amendments to the grand narrative are now technically possible.

Writing in *Wired* magazine in July 2013, Harvard computer-science professor Jonathan Zittrain described this as "a worrisome trend" and called for digital books and other texts to be placed under the control of readers and libraries—presumed to have a vested interest in the sanctity of text—rather than with distributors and digital gatekeepers. Most insidious of all, Zittrain noted, was the fact that changes can be made with no evidence that things were ever any other way. "If we're going to alter or destroy the past," he wrote, "we should [at least] have to see, hear and smell the paper burning."[42]

A Standardized Taste

In the early 1980s, a computer science and electronic engineering graduate from UC Berkeley set out to create a musical synthesizer. What Dave Smith wanted was to establish a standardized protocol for communication between the different electronic musical instruments made by different manufacturers around the world. What he came up with was christened the "Musical Instrument Digital Interface" and—better known by the name MIDI—became the entrenched unitary measurement for music. As a musical medium, MIDI is far from perfect. Although it can be used to mimic a wide palette of sounds using a single keyboard, it retains the keyboard's staccato, mosaic qualities, which means that it cannot emulate the type of curvaceous sounds produceable by, say, a talented singer or saxophonist. As virtual-reality innovator (and talented musician) Jaron Lanier observes:

> Before MIDI, a musical note was a bottomless idea that transcended absolute definition . . . After MIDI, a musical note [is] no longer just an idea, but a rigid, mandatory structure you couldn't avoid in the aspects of life that had gone digital.[43]

This sort of technological "lock-in" is an unavoidable part of measurement. The moment we create a unitary standard, we also create limitations. More than two centuries before Dave Smith created MIDI, an 18th-century Scottish philosopher named David Hume wrote an essay entitled "(Of the) Standard of Taste." In it, Hume argued that the key component to art (the thing that would come after the equals sign were it formulated as an equation) was the presence of what he termed "agreeableness." Hume observed, "it is natural for us to seek a Standard of Taste; a rule, by which the various sentiments of men may be reconciled."[44]

Unlike many of the figures discussed at the start of this chapter, Hume believed that there were not objective measures of aesthetic value, but that these were rather subjective judgments. As he phrased it, "to seek the real beauty, or the real deformity, is as fruitless an enquiry, as to seek the real sweet or real bitter." At the same time, Hume acknowledged that, within subjectivity, aspects do indeed exist that are either "calculated to please" or "displease"—thus bringing about his "standard of taste."

Hume was ahead of his time in various ways. In recent years, a number of organizations around the world have been investigating what is referred to as "Emotional Optimization." Emotional Optimization relates to the discovery that certain parts of the brain correspond to different emotions. By asking test subjects to wear electroencephalography (EEG) brain caps, neuroscientists can measure the electrical activity that results from ionic current flows within the neurons of the brain. These readings can then be used to uncover the positive and negative reactions experienced by a person as they listen to a piece of music or watch a scene from a film. Through the addition of machine-learning tools, the possibility of discovering which low-level features in art prompt particular emotional responses becomes a reality.

Looking to the future, the potential of such work is clear. The addition of a feedback loop, for instance, would allow users not simply to have their EEG response to particular works read, but also to dictate the mood they wanted to achieve. Instead of having playlists to match our mood, a person would have the option of entering their desired emotion into a computer, with a customized playlist then generated to

provoke that specific response. This may have particular application in the therapeutic world to help treat those suffering
from stress or forms of depression. Runners, meanwhile, could
have their pulse rates measured by the headphones they're
wearing, with music selected according to whether heart rate
rises or falls. Translated to literature, electronic novels could
monitor the electrical activity of neurons in the brain while
they are being read, leading to algorithms rewriting sections
to match the reactions elicited. In the same way that a stand-up
comic or live musician subtly alters their performance to fit a
particular audience, so too will media increasingly resemble its
consumer. The medium might stay the same, but the message
will change depending on who is listening.

In an article published in the *New Statesman*, journalist
Alexandra Coughlan refers to this idea as "aural pill-popping,"
in which Emotional Optimization will mean that there will be
"one [music] track to bring us up [and] another to bring us
down."[45] This comment demonstrates a belief in functional
form—the idea that, as I described earlier in this chapter, it is
desirable that art be "made useful" in some way. Coughlan's suggestion of "aural pill-popping" raises a number of questions—
not least whether the value of art is simply as a creative substitute
for mind-altering drugs.

We might feel calm looking at Mark Rothko's *Untitled
(Green on Blue)* painting, for example, but does this relegate it
to the artistic equivalent of Valium? In his book *To Save Everything, Click Here*, Belarusian technology scholar Evgeny
Morozov takes this utilitarian idea to task. Suppose, Morozov
says, that Google (selecting one company that has made clear
its ambitions to quantify everything) knows that we are not at

our happiest after receiving a sad phone call from an ex-girlfriend. If art equals pleasure—and the quickest way to achieve pleasure is to look at a great painting—then Google knows that what we need more than anything for a quick pick-me-up is to see a painting by Impressionist painter Renoir:

> Well, Google doesn't exactly "know" it; it knows only that you are missing 124 units of "art" and that, according to Google's own measurement system, Renoir's paintings happen to average in the 120s. You see the picture and—boom!—your mood stays intact.[46]

Morozov continues his line of inquiry by asking the pertinent questions that arise with such a proposition. Would keeping our mood levels stabilized by looking at the paintings of Renoir turn us into a world of art lovers? Would it expand our horizons? Or would such attempts to consume art in the manner of self-help literature only serve to demean artistic endeavors? Still more problems not touched on by Morozov surface with efforts to quantify art as unitary measures of pleasure, in the manner of Sergei Eisenstein's "attractions." If we accept that Renoir's work gives us a happiness boost of, say, 122, while Pablo Picasso's score languishes at a mere 98, why bother with Picasso's work at all?

Similarly, let's imagine for a moment that the complexity of Beethoven's *7th Symphony* turns out to produce measurably greater neurological highs than Justin Bieber's hit song "Baby," thereby giving us the ability to draw a mathematical distinction between the fields of "high" and "low" art. Should this prove to be the case, could we receive the same dosage of artistic nourishment—albeit in a less efficient time frame—by watching multiple episodes of *Friends* (assuming the sitcom is classified as "low" art) as we could from reading Leo Tolstoy's *War and Peace* (supposing that it is classified as "high" art)? Ultimately, presuming that *War and Peace* is superior to *Friends*, or that Beethoven is superior to Justin Bieber, simply because they top up our artistic needs at a greater rate of knots, is essentially the same argument as suggesting that James Patterson is a greater novelist than J. M. Coetzee on the basis that data gathered by Kindle shows that Patterson's *Kill Alex Cross* can be read in a single afternoon, while Coetzee's *Life & Times of Michael K* takes several days, or even weeks. It may look mathematically rigorous, but something doesn't quite add up.

The Dehumanization of Art

All of this brings us ever closer to the inevitable question of whether algorithms will ever be able to generate their own art. Perhaps unsurprisingly, this is a line of inquiry that provokes heated comments on both sides. "It's only a matter of *when* it happens—not if," says Lior Shamir, who built the automated art critic I described earlier. Much as Epagogix's movie prediction system spots places in a script where a potential yield is not where it should be and then makes recommendations accordingly, so Shamir is convinced that in the long term his creation will be able to spot the features great works of art have in common and generate entire new works accordingly.

While this might seem a new concept, it is not. In 1787, Mozart anonymously published what is referred to in German as *Musikalisches Würfelspiel* ("musical dice game"). His idea was simple: to enable readers to compose German waltzes, "without the least knowledge of music . . . by throwing a certain number with two dice." Mozart provided 176 bars of music, arranged in 16 columns, with 11 bars to each column. To select the first musical bar, readers would throw two dice and then choose the corresponding bar from the available options. The technique was repeated for the second column, then the third, and so on. The total number of possible compositions was an astonishing $46 \times 1,000,000,000,000,000$, with each generated work sounding Mozartian in style.[47]

A similar concept—albeit in a different medium—is the current work of Celestino Soddu, a contemporary Italian architect and designer who uses what are referred to as "genetic algorithms" to generate endless variations on individual themes. A genetic algorithm replicates evolution inside a computer, adopting the idea that living organisms are the consummate problem solvers and using this to optimize specific solutions. By inputting what he considers to be the "rules" that define, say, a chair or a Baroque cathedral, Soddu is able to use his algorithm to conceptualize what a particular object might look like were it a living entity undergoing thousands of years of natural selection. Because there is (realistically speaking) no limit to the amount of results the genetic algorithm can generate, Soddu's "idea-products" mean that a trendy advertising agency could conceivably fill its offices with hundreds of chairs, each one subtly different, while a company engaged in building its new corporate headquarters might generate thousands of separate designs before deciding upon one to go ahead with.

There are, however, still problems with the concept of creating art by algorithm. Theodor Adorno and Max Horkheimer noted in the 1940s how formulaic art does not offer new experiences, but rather remixed versions of what came before. Instead of the joy of being exposed to something new, Adorno saw mass culture's reward coming in the form of the smart audience member who "can guess what is coming and feel flattered when it does come."[48] This prescient comment is backed up by algorithms that predict the future by establishing what has worked in the past. An artwork in this sense might achieve a quantifiable perfection, but it will only ever be perfection measured against what has already occurred.

For instance, Nick Meaney acknowledges that Epagogix would have been unable to predict the huge success of a film like *Avatar*. The reason: there had been no $2 billion films before to measure it against. This doesn't mean that Epagogix wouldn't have realized it had a hit on its hands, of course. "Would we have said that it would earn what it did in the United States? Probably not," Meaney says. "It would have been flagged up as being off the scale, but because it was off the scale there was nothing to measure it against. The next *Avatar*, on the other hand? Now there's something to measure it against."

The issue becomes more pressing when it relates to the generating of new art, rather than the measurement of existing works. Because Lior Shamir's automated art critic algorithm measures works based on 4,024 different numerical descriptors, there is a chance that it might be able to quantify what would comprise the best illustration of, say, pop art and generate an artwork that conforms to all of these criteria. But these criteria are themselves based upon human creativity. Would it be possible for algorithms themselves to move art forward in a meaningful way, rather than simply aping the style of previous works? "At first, no," Shamir says. "Ultimately, I would be very careful in saying there are things that machines can not do."

A better question might be whether we would accept such works if they did—knowing that a machine rather than a human artist had created them? For those that see creativity as a profoundly human activity (a relatively new idea, as it happens), the question is one that goes beyond technical ability and touches on somewhat close to the essence of humanity.

In 2012, the London Symphony Orchestra took to the stage to perform compositions written entirely by a music-generating algorithm called Iamus.[49] Iamus was the project of professor and entrepreneur Francisco Vico, under whose coding it has composed more than one billion songs across a wide range of genres. In the aftermath of Iamus's concert, a staff writer for the *Columbia Spectator* named David Ecker put pen to paper (or rather finger to keyboard) to write a polemic taking aim at the new technology. "I use computers for damn near everything, [but] there's something about this computer that I find deeply troubling," Ecker wrote.

> I'm not a purist by any stretch. I hate overt music categorization, and I hate most debates about "real" versus "fake" art, but that's not what this is about. This is about the very essence of humanity. Computers can compete and win at *Jeopardy!*, beat chess masters, and connect us with people on the other side of the world. When it comes to emotion,

however, they lack much of the necessary equipment. We live every day under the pretense that what we do carries a certain weight, partly due to the knowledge of our own mortality, and this always comes through in truly great music. Iamus has neither mortality nor the urgency that comes with it. It can create sounds—some of which may be pleasing—but it can never achieve the emotional complexity and creative innovation of a musician or a composer. One could say that Iamus could be an ideal tool for creating meaningless top-40 tracks, but for me, this too would be troubling. Even the most transient and superficial of pop tracks take root in the human experience, and I believe that even those are worth protecting from Iamus.[50]

Perhaps there is still hope for those who dream of an algorithm creating art. However, as Iamus's Francisco Vico points out: "I received one comment from a woman who admitted that Iamus was a milestone in technology. But she also said that she had to stop listening to it, because it was making her *feel* things. In some senses we see this as creepy, and I can fully understand that. We are not ready for it. Part of us still thinks that computers are Terminators that want to kill us, or else simple tools that are supposed to help us with processing information. The idea that they can be artists, too, is something unexpected. It's something new."

CONCLUSION

Predicting the Future

In 1954, a 34-year-old American psychology professor named Paul E. Meehl published a groundbreaking book with a somewhat unwieldy title. *Clinical vs. Statistical Prediction: A Theoretical Analysis and a Review of the Evidence* presented 20 case studies in which predictions made by statistical algorithms were compared with clinical predictions made by trained experts.[1]

A sample study asked trained counselors to predict the end-of-year grades of first-year students. The counselors were allowed three-quarters of an hour to interview each student, along with access to their previous grades, multiple aptitude tests, and a personal statement that ran four pages in length. The algorithm, meanwhile, required only high school grades and a single aptitude test. In 11 out of 14 cases, the algorithm proved more accurate at predicting students' finishing grades than did the counselors.

The same proved true of the textbook's other studies, which

analyzed topics as diverse as parole violation rates (as per Chapter 3's Richard Berk) to would-be pilots' success during training. In 19 out of 20 cases—which Meehl later argued should be modified to a clean sweep—the statistical algorithms were demonstrably more accurate than those made by the experts, and almost always required less data in order to get there. "It is clear," Meehl concluded, "that the dogmatic, complacent assertion sometimes heard from clinicians that 'naturally' clinical predictions, being based on 'real understanding' is superior, is simply not justified by the facts to date."

Meehl was, perhaps understandably, something of an outsider in academic circles from this point on. His anti-expert stance amounted to suggesting—in the words of a colleague quoted in Meehl's 2003 *New York Times* obituary—that "clinicians could be replaced by a clerk with a hand-cranked Monroe calculator."[2] (Meehl's status as prototypical Internet troll was only further added to by the publishing of a later paper in his career, entitled "Why I Do Not Attend Case Conferences," in which he dismissed academic conferences on the basis that they were boring to the point of offensiveness.)

Regardless of his divisive status at the time, Meehl's views of the predictive power of algorithms have been borne out in the years since. In the roughly 200 similar studies that have been carried out in the half century since, algorithms have triumphed over human intuition with a success rate of around 60 percent. In the remaining 40 percent, the difference between statistical and clinical predictions proved statistically insignificant, still representing a tick in the "win" column for the algorithmic approach, since this is almost always cheaper than hiring an expert.

The Power of Thinking Without Thinking

Why do algorithms interest us? The first point to make is that it is quite likely that many of the computer scientists reading this book will be the same people who would have picked up a similar book in 1984, or 1964. But not all of us (including this writer) are computer scientists by trade, and the question of how and why a once obscure mathematical concept came to occupy the front page of major newspapers and other publications was one that often occurred to me when I was carrying out my research.

In this final chapter, I would like to share some of my thoughts on that question. It seems obvious to point out that the reason for this comes down to the growing role that algorithms have to play in all of our lives on a daily basis. Search engines like Google help us to navigate massive databases of information. Recommender systems like those employed by Amazon meanwhile map our preferences against those of other people and suggest new bits of culture for us to experience. On social networking sites, algorithms highlight news that is "relevant" to us, and on dating sites like eHarmony they match us up with potential life partners. It is not "cyberbole," then, to suggest that algorithms represent a crucial force in our participation in public life.

They go further than the four main areas I have chosen to look at in this book, too. For instance, algorithmic trading now represents a whopping 70 percent of the U.S. equity market, running on supercomputers that are able to buy and sell millions of shares at practically the speed of light. Algorithmic trading has become a race measured in milliseconds, with billions of dollars dependent on the laying of new fiber-optic cables that will shave just five milliseconds off the communication time between financial markets in London and New York. (To put this in perspective, it takes a human 300 milliseconds to blink.)[3]

Medicine, too, has taken an algorithmic turn, as doctors working in hospitals are often asked to rely on algorithms rather than their own clinical judgment. In his book *Blink: The Power of Thinking Without Thinking*, Malcolm Gladwell recounts the story of one hospital that adopted an algorithm for diagnosing chest pain. "They instructed their doctors to gather less information on their patients," Gladwell writes, explaining how doctors were told to instead zero in "on just a few critical pieces of information about patients . . . like blood pressure and the ECG—while ignoring everything else, like the patient's age and weight and medical history. And what happened? Cook County is now one of the best places in the United States at diagnosing chest pain."[4] Recent medical algorithms have been shown to yield equally impressive results in other areas, such as an algorithm able to diagnose for Parkinson's disease by listening to a person's voice over the telephone, and another pattern-recognition algorithm able to, quite literally, "sniff" for diseases like cancer.

Algorithmizing the World

Can everything be subject to algorithmization? There are two ways to answer this question. The first is to approach it purely on a technical level. At present, no, everything cannot be "solved" by an algorithm. At time of writing, for instance, recognizing objects with anything close to the ability of a human is still a massive challenge. A young child only has to be shown a handful of "training examples" in order to identify a particular object—even if they have never seen that object before. An algorithm designed for a similar task, however, will frequently require long practice sessions in which the computer is shown thousands of different versions of the same thing and corrected by a human when it is wrong. Even then, an algorithm may struggle with the task when carrying it out in a real-world environment, in which it is necessary to disambiguate contours belonging to different overlapping objects.

Computer scientist and teacher John MacCormick similarly gives the example of an algorithm's unsuitability for being used as a teaching aid for grading students' work, since this is a task that is too complex (and, depending on the subject, too subjective) for a bot to carry out.[5] Could both of these tasks be performed algorithmically in the future as computers continue to get more powerful? Absolutely. It is for this reason that it is dangerous to bet against a bot. A decade ago, respected MIT and Harvard economists Frank Levy and Richard Murnane published a well-researched book entitled *The New Division of Labor*, in which they compared the respective capabilities of human workers and computers. In an optimistic second chapter called "Why People Still Matter," the authors described a spectrum of information-processing tasks ranging from those that could be handled by a computer (e.g., arithmetic), to those that only a human could do. One illustration they gave was that of a long-distance truck driver:

The . . . truck driver is processing a constant stream of [visual, aural and tactile] information from his environment. . . . [T]o program this behavior we could begin with a video camera and other sensors to capture the sensory input. But executing a left turn against oncoming traffic involves so many factors that it is hard to imagine discovering the set of rules that can replicate a driver's behavior. . . . Articulating [human] knowledge and embedding it in software for all but highly structured situations are at present enormously difficult tasks. . . . Computers cannot easily substitute for humans in [jobs like truck driving].[6]

At least one part of this assertion is correct: computers cannot *easily* substitute for humans when it comes to driving. Certainly Levy and Murnane were not mistaken at the time that they were writing. The year their book was released, DARPA announced its Grand Challenge, in which entrants from the country's top AI laboratories competed for a $1 million prize by constructing driverless vehicles capable of navigating a 142-mile route through the Mojave Desert. The "winning" team made it less than eight miles (in several hours) before it caught fire and shuddered to a halt.

A lot can change in a decade, however, as you will know from reading Chapter 3, in which I discuss the success of Google's self-driving cars. Should such technologies prove suitably efficient, there is every possibility that they will take over the jobs currently occupied by taxi drivers and long-distance drivers.

Similar paradigm shifts are now taking place across a wide range of fields and industries. Consider Amazon, for instance. In Amazon's early days (when it was just an online book retailer, rather than the leviathanic "everything store" it was referred to as in a recent biography) it featured two rival departments, whose interdepartmental squabbling serves as a microcosm of sorts for the type of fight regularly seen in the age of The Formula. One department was made up of the editorial staff, whose job it was to review books, write copy for the website's home page and provide a reassuringly human voice to customers still wary about handing over their credit card details to a faceless machine. The other group was referred to as the personalization team and was tasked with the creation of algorithms that would recommend products to individual users.

Of the two departments, it was this latter division that won both the short-term battle and the long-term war. Their winning weapon was called Amabot and replaced what had previously been person-to-person, handcrafted sections of Amazon's website with automatically generated suggestions that conformed to a standardized layout. "The system handily won a series of tests and demonstrated it could sell as many products as the human editors," wrote Brad Stone in his well-researched 2013 history of Amazon.

After the editorial staff had been rounded up and either laid off or else assigned to other parts of the company, an employee summed up the prevailing mood by placing a "lonely hearts" advertisement in the pages of a local Seattle newspaper on Valentine's Day in 2002, addressing the algorithm that had rendered them obsolete:

DEAREST AMABOT: If you only had a heart to absorb our hatred . . . Thanks for nothing, you jury-rigged rust bucket. The gorgeous messiness of flesh and blood will prevail![7]

This is a sentiment that is still widely argued—particularly when algorithms take on the kind of humanities-oriented fields I have approached in this book. However, it is also necessary to note that drawing a definite line only capable of being crossed by the "gorgeous messiness of [the] flesh" is a little like Levy and Murnane's statements about which jobs are safe from automation.

Certainly, there are plenty of jobs and other areas of life now carried out by algorithm, which were previously thought to have been the sole domain of humans. Facial recognition, for instance, was once considered to be a trait performable only by a select few higher-performing animals—humans among them. Today algorithms employed by Facebook and Google regularly recognize individual faces out of the billions of personal images uploaded by users.

Much the same is true of language and automated translation. "There is no immediate or predictable prospect of useful machine translation," concluded a U.S. National Academy of Sciences committee in 1965. Leap forward half a century and Google Translate is used on a daily basis, offering two-way translation between 58 different languages: 3,306 separate translation services in all. "The service that Google provides appears to flatten and diversify inter-language relations beyond the wildest dreams of even the E.U.'s most enthusiastic language parity proponents," writes David Bellos, author of *Is That a Fish in Your Ear?: Translation and the Meaning of Everything*.[8] Even if Google Translate's results aren't always perfect, they are often "good enough" to be useful—and are getting better all the time.

The Great Restructuring

What is notable about The Formula is how, in many cases, an algorithm can replace large numbers of human workers. Jaron Lanier makes this point in his most recent book, *Who Owns The Future?*, by comparing the photography company Kodak with the online video-sharing social network Instagram. "At the height of its power . . . Kodak employed more than 140,000 people and was worth $28 billion," Lanier observes. "They even invented the first digital camera. But today Kodak is bankrupt, and the new face of digital photography has become Instagram. When Instagram was sold to Facebook for $1 billion in 2012, it employed only 13 people. Where did all those jobs disappear to? And what happened to the wealth that all those middle-class jobs created?"[9]

What causes shock for many people commenting on this subject is how indiscriminate the automation is. Nothing, it seems, is safe from a few well-designed algorithms offering speed, efficiency and value for money. Increasing numbers of books carry doom scenarios related to industries struggling in the wake of The Formula. In *Failing Law Schools*, law professor Brian Tamanha points to U.S. government statistics suggesting that until 2018 there will only be 25,000 new openings available for young lawyers—despite the fact that law schools will produce around 45,000 graduates during that same time frame. It is quite possible, Tamanha writes, that this ratio will one day be remembered as the "good old days." Indeed, it is quite conceivable to imagine a future in which law firms stop hiring junior and trainee lawyers altogether, and pass much of this work over to artificial intelligence systems instead. In keeping with this, a number of experts predict that there will be between 10 and 40 percent fewer lawyers a decade from now as there are today.[10]

As Erik Brynjolfsson and Andrew McAfee suggest in their pamphlet "Race Against the Machine," this is not so much the result of a Great Recession or a Great Stagnation, so much as it is a Great Restructuring.[11] The new barometer for which jobs are safe from The Formula has less to do with the social class of those traditionally holding them than it does to do with a trade-off between cost and efficiency. Professions and fields that have evolved to operate as inefficiently as possible (lawyers, accountants, barristers and legislators, for example) while also charging the most money will be particularly vulnerable when it comes to automation. To survive—as economist Theodore Levitt famously phrased it in his 1960 article "Marketing Myopia"—every industry must "plot the obsolescence of what now produces their livelihood."[12]

In the new algorithmic world, it is the computer scientists and mathematicians who will be increasingly responsible for making cultural determinations and will ultimately thrive in the Great Restructuring. Others will suffer the "end of work" described by social theorist Jeremy Rifkin in his book of the same name. This is a workplace in which "fewer and fewer workers will be needed to produce the goods and services for the global population." As costs of everything from legal bills to entertainment come down, so too will the availability of many types of work decrease. As André Gorz writes in *Farewell to the Working Class*, "The majority of the population [will end up belonging to] the post-industrial neo-proletariat which, with no job security or definite class identity, fills the area of probationary, contracted, casual, temporary and part-time employment."[13] As job security and class identity are replaced by automation and algorithmic user profiles, the world may finally get the "twenty-hour working week and retirement at fifty" that previous generations of techo-utopianists dreamed about. It just won't necessarily be voluntary.[14]

Spoons Instead of Shovels

There is a famous anecdote about the American economist and statistician Milton Friedman visiting a country in Asia during the 1960s. Taken to a worksite where a new canal was being excavated, Friedman was shocked to see that the workers were using shovels instead of modern tractors and earthmovers. "Why are there so few machines?" he asked the government bureaucrat traveling with him. "You don't understand," came the answer. "This is a jobs program." Friedman considered for a second, then replied, "Oh, I thought you were trying to build a canal. If it's jobs you want, then you should give these workers spoons, not shovels."

You could, of course, extend this to any number of technologies. Tractors are more efficient earthmovers than shovels, as shovels are more efficient than spoons, and spoons are more efficient than hands. The question is, where do we stop the process? The cultural theorist Paul Virilio once pointed out how the invention of the ship was also the invention of the shipwreck. If this is the case, then how many shipwrecks do

we need before we stop building ships? Those looking for stories of algorithms run amok can certainly find them with relative ease. On May 6, 2010, the Dow Jones Industrial Average
plunged 1,000 points in just 300 seconds—effectively wiping
out close to $1 trillion of wealth in a stock market debacle that
became known as the Flash Crash. Unexplained to this day,
the Flash Crash has been pinned on everything from the
impact of high-speed trading to a technical glitch.[15]

Yet few people would seriously put forward the view that
algorithms are, in themselves, bad. Indeed, it's not simply a
matter of algorithms doing the jobs that were once carried out
manually; in many cases algorithms perform tasks that would
be impossible for a human to perform. Particularly, algorithms
like those utilized by Google that rely on unimaginably large
datasets could never be reenacted by hand. Consider also the
algorithm developed by mathematician Max Little that is able
to diagnose Parkinson's disease down the phone line, by "listening" to callers' speech patterns for vocal tremors that are
inaudible to the human ear.[16]

The French economist Jean Fourastié humorously asked
whether prehistoric man felt the same trepidation at the invention of the bronze sword that 20th-century man did at the
birth of the atomic bomb. As technologies are invented and
prove not to be the end of humanity, they recede into background noise, where they become fodder for further generations of disruptive technology, just as the strongest lions
eventually weaken and are overtaken by younger, fitter ones.
Confusing the matter further is the complex relationship we
enjoy with technology on a daily basis. Like David Ecker, the
Columbia Spectator journalist I quoted in the last chapter,

most of us hold concerns over "bad" uses of technology, while enjoying everything good technology makes possible. To put it another way, how did I find out the exact details of the Flash Crash I mentioned above? I Googled it.

Objectivity in the Post-mechanical Age

One topic I continued to butt up against during the writing of this book (and in my other tech writing for publications like *Fast Company* and *Wired*) is the subject of objectivity. In each chapter of this book, the subject of objectivity never strayed far from either my own mind or the various conversations I enjoyed with the technologists I had the opportunity to interview. In Chapter 1, the question of objectivity arose with the idea that there are certain objective truths algorithms can ascertain about an individual, be those the clinician-free means by which Larry Smarr diagnosed himself with Crohn's disease or the "dividual" profiles constructed by companies like Quantcast. In Chapter 2, the dream of objectivity revolved around the ultrarational matching process at the heart of algorithmic dating, supposed to provide us with a "better" way of matching us with romantic partners. In Chapter 3, objectivity was a way of making the law fairer, by creating legal algorithms that would ensure that the law was enforced the same way every time. Finally, in Chapter 4, objectivity was about the universal rules relating to what defines something as a work of art.

Objectivity is a term that is often associated with algorithms, and companies in thrall of them. For example, Google notes in its "Ten Things We Know to Be True" manifesto that "our users trust our objectivity." Were we to think for too long about the fact that we expect a publicly traded company to be entirely objective (or even that such a thing is possible when it comes to filtering and ranking information), we might see the fundamental flaw in this conundrum—but then again, when Google is providing search results in 0.21 seconds, there isn't a great deal of time to think. "This is a very unique way of thinking about the world," says scholar Ted Striphas, author of *The Late Age of Print*, who has spent the past several years investigating what he calls algorithmic culture. "It's about degrees of pure objectivity, where you are never in the realm of subjectivity; you're only in the realm of getting closer and closer to some inexorable truth . . . You are never wrong, you're only ever *more* right."

As it happens, Google may actually be telling the truth here: their users really *do* seem to trust in their objectivity. According to a survey of web users carried out in 2005, only 19 percent of individuals expressed a lack of trust in their search engines, while more than 68 percent considered the search engines they used regularly to be fair and unbiased.[17]

Science-fiction author Arthur C. Clarke famously wrote, "any sufficiently advanced technology is indistinguishable from magic." Just like photography appeared to people over 100 years ago, so too does the speed with which algorithms work make us view them as authoritative and unaffected by human fallibility. Paste a block of text into Google's translation services and in less than a second its algorithms can transform the words into any one of 58 different languages. The same is true of Google's search algorithms, which, as a result of its "knowledge" about individual users, allow our specific desires and requirements to be predicted with an almost preternatural ability. "Google works

for us because it seems to read our minds," says media scholar
Siva Vaidhyanathan, "and, in a way, it does."[18]

As with magic, our reverence for Google's work comes partly
because we see only the end result and none of the working. Not
only are these black-boxed and obscured, they are practically
instantaneous. This effect doesn't only serve to fool the technologically uniformed. At the start of his book *Nine Algorithms
That Changed the Future*, accomplished mathematician and
computer scientist John MacCormick writes, "at the heart of
every algorithm . . . is an ingenious trick that makes the whole
thing work." He goes on to expand upon the statement, suggesting that:

> Since I'll be using the word "trick" a great deal, I should
> point out that I'm not talking about the kind of tricks that
> are mean and deceitful—the kind of trick a child might
> play on a younger brother or sister. Instead, the tricks . . .
> resemble tricks of the trade or even magic tricks: clever
> techniques for accomplishing goals that would otherwise
> be difficult or impossible.[19]

While perhaps a well-intentioned distinction, MacCormick's
error is his casual assumption about a sense of algorithmic
morality. Stating that algorithms and the goals they aim to
accomplish are neither good nor bad (although, to return to
Melvin Kranzberg's first law of technology, nor are they neutral) seems an extraordinarily sweeping and unqualified statement. Nonetheless, it is one that has been made by a number
of renowned technology writers. In a controversial 2008 art-

icle published in *Wired* magazine, journalist Chris Anderson announced that the age of big datasets and algorithms equaled what he grandly referred to as "The End of Theory." No more would we have to worry about the elitist theories of so-called experts, Anderson said.

> There is now a better way. Petabytes allow us to say: "Correlation is enough." We can stop looking for models. We can analyze the data without hypotheses about what it might show. We can throw the numbers into the biggest computing clusters the world has ever seen and let statistical algorithms find patterns where science cannot.[20]

The problem with Anderson's enthusiastic embrace of The Formula, of course, is his failure to realize that data mining, even on large datasets, is itself founded on a theory. As I have shown throughout this book, algorithms can often reflect the biases of their creators—based upon what it is that they deem to be important when answering a particular question. When an algorithm is created to determine what information is relevant, or the best way to solve a problem, this represents a hypothesis in itself. Even data is not free of human bias, from what data is collected to the manner in which it is cleaned up and made algorithm-ready. A computational process that seeks to sort, classify and create hierarchies in and around people, places, objects and ideas carries considerable political connotations. So too does a subject like the kind of user categorization I discussed in Chapter 1. What are the categories? Who belongs to each category? And how do we know that categories are there to help—rather than hinder—us?

Can an Algorithm Defame?

In March 2013, a T-shirt manufacturer called Solid Gold Bomb found itself in a heated row with Amazon over a slogan generated by an algorithm. Seizing upon the trend for "Keep Calm and . . ." paraphernalia sweeping the UK at the time, Solid Gold Bomb created a simple bot to generate similar designs by running through an English dictionary and randomly matching verbs with adjectives. In all, 529,493 similarly themed clothing items appeared on Amazon's site—with the T-shirts only printed once a customer had actually bought one. As website *BoingBoing* wrote of the business plan: "It costs [Solid Gold Bomb] nothing to create the design, nothing to submit it to Amazon and nothing for Amazon to host the product. If no one buys it then the total cost of the experiment is effectively zero. But if the algorithm stumbles upon something special, something that is both unique and funny and actually sells, then everyone makes money."[21]

Unfortunately, no one at Solid Gold Bomb had apparently considered the possibility that the algorithm might generate offensive slogans while running down its available list of words. The variation that set Amazon's blood boiling was the offensive "Keep Calm and Rape a Lot"—although had this not done the job, it would likely have been equally appalled by the misogynistic "Keep Calm and Hit Her," or the ever-unpopular "Keep Calm and Grope On." When it found out what was going on, Amazon responded by immediately pulling all of Solid Gold Bomb's online inventory. The T-shirt manufacturer (which wound up going out of business several months later) was upset. Why, its owners wondered, should they be punished when the fault was not with any human agency—but with an algorithm, for whom the words in question meant nothing?

A similar variation on this problem took place the previous year, when Google's "auto-complete" algorithm came under fire for alleged defamation from Bettina Wulff, wife of the former German president Christian Wulff.[22] Originally an algorithm designed to help people with physical disabilities increase their typing speed, auto-complete was added to Google's functionality as a way of saving users time by predicting their search terms before they had finished typing them. "Using billions of searches, Google has prototyped an anonymous profile of its users," says creator Marius B. Well. "This reflects the fears, inquiries, preoccupations, obsessions and fixations of the human being at a certain age and our evolution through life."[23] Type Barack Obama's name into the Google search box, for example, and you would be presented with potentially useful suggestions including "Barack Obama," "Barack Obama Twitter," "Barack Obama quotes" and "Barack Obama facts." Type in the name of United Kingdom deputy prime minister Nick Clegg, on the other hand, and you are liable to find "Nick Clegg is a prick," "Nick Clegg is a liar," "Nick Clegg is sad" and "Nick Clegg is finished." Of these two camps, Bettina Wulff's suggested searches fell more in line with Nick Clegg's than Barack Obama's. A person searching for Wulff's name was likely to find search terms linking her to prostitution and escort businesses.[24]

Realizing the effect that this was likely to have on someone searching for her name online, Wulff took Google to court and won. A German court decided that Google would have to ensure that the terms algorithmically generated by auto-complete were not offensive or defamatory in any way. Google was upset, claiming to be extremely "disappointed" by the ruling, since this impugned the supposed objective impartiality of its algorithms. "We believe that Google should not be held liable for terms that appear in auto-complete as these are predicted by computer algorithms based on searches from previous users, not by Google itself," said a spokesperson for the company. "We are currently reviewing our options." The problem is the amount, both figuratively and literally, that Google has fetishistically invested in its algorithmic vision. Like the concept of sci-fi author Philip K. Dick's "minority reports" referenced in Chapter 3, if the algorithm proves fallible then tugging on this thread could have catastrophic results. "Google's spiritual deferral to 'algorithmic neutrality' betrays the company's growing unease with being the world's most important information gatekeeper," writes Evgeny Morozov in his book *The Net Delusion*. "Its founders prefer to treat technology as an autonomous and fully objective force rather than spending sleepless nights worrying about inherent biases in how their systems—systems that have grown so complex that no Google engineer fully understands them—operate."[25]

A narrative thread often explored in books and articles about Google is the degree to which Google's rise has helped speed up the decline of traditional news outlets, like newspapers. In this sense, Google has displaced traditional media, even though it does not generate news stories itself. If Google's algorithms ought to be subject to the same standards as newspapers, though, this poses some problems. In a classic study of newsroom objectivity, sociologist Gaye Tuchman observed that it was a fear of defamation that kept journalism objective. By reporting the views of others rather than relying on their own opinion, journalists were protected against allegations that they were biased in their reporting. In terms of Google's auto-complete algorithm, it had also relied on quoting others rather than expressing opinions, since its suggested terms were based on the previous searches of thousands, or even millions, of users. By not censoring these searches, however, and keeping its algorithms apparently objective, Google had been accused of defamation.[26]

Why Is Twitter Like a Newspaper? (And Why Isn't Google?)

The Bettina Wulff case marked several interesting developments. For one thing, it represented one of the first times that the politics of algorithms became the subject of a legal trial. Algorithms can be difficult to criticize. In contrast to the likes of a service such as Google Street View—whose arrival sparked street protests in some countries due to its perceived violation of privacy—the invisibility of an algorithm can make it tough to spot its effects. It is one thing to be able to log on to a server and see a detailed image of your house as taken from the end of your driveway. It is another to critique the inner workings of the algorithms that underpin companies such as Google, Facebook and Amazon. In the majority of cases, these algorithms are black-boxed in such a way that users have no idea how they work. Like the changing concept of transparency in the digital world (something I discussed in Chapter 3), often the idea that complex technology can work under a purposely simplistic interface is viewed by Silicon Valley decision-makers as a key selling point. Speaking about Google's algorithms in 2008, Marissa Mayer—today the president and CEO of Yahoo!—had the following to say:

> We think that that's the best way to do things. Our users don't need to understand how complicated the technology and the development work that happens behind this is. What they do need to understand is that they can just go to a box, type what they want, and get answers.[27]

Wulff's concern about the politics of auto-correct was also proof positive of the power algorithms carry in today's world in terms of changing the course of public opinion. By suggesting particular unflattering searches, a user with no preconceived views about Bettina Wulff could be channeled down a particular route. In this way, algorithms aren't just predicting user behavior—they are helping dictate it. Consider, for example, Netflix's recommendation algorithms, which I discussed in Chapter 4. Netflix claims that 60 percent of its rentals are done according to its algorithm's suggestions, rather than users specifically searching for a title. If this is the case, are we to assume that the algorithm simply guessed what users would next want to search for, or that the users in fact made a certain selection because an algorithm had placed particular options in front of them?

Here the question becomes almost irrelevant. As the sociologists William and Dorothy Thomas famously noted, "If men define situations as real, they are real in their consequences." Or to put it in the words Kevin Slavin memorably used during his TED Talk, "How Algorithms Shape Our World," the math involved in such computer processes has transitioned from "something that we extract and derive from the world, to something that actually starts to shape it."[28]

This can quite literally be the case. On September 6, 2008, an algorithm came dangerously close to driving United Airlines' parent company UAL out of business. The problem started when a reporter for a news company called Income Securities Advisors entered the words "bankruptcy 2008" in Google's search bar and hit "enter." Google News immediately pointed the reporter to an article from the *South Florida Sun-Sentinel*, revealing that UAL had filed for bankruptcy protection. The reporter—who worked for a company responsible for feeding stories to the powerful Bloomberg news service—immediately posted an alert to the news network, lacking any further contextual information, entitled "United Airlines files for Ch. 11 to cut costs." The news that the airline was seeking legal protection for its debtors was quickly read by thousands of influential readers of Bloomberg's financial news updates. The problem, as later came to light, was that the news was not actually new, but referred to a bankruptcy filing from 2002, which the company had later successfully emerged from, thanks to a 2006 reorganization. Because the *South Florida Sun-Sentinel* failed to list a date with its original news bulletin, Google's algorithms had assigned it one based upon the September 2008 date that its web-crawling software found and indexed the article. As a result of the misinformation, UAL stock trading on the NASDAQ plummeted from $12.17 per share to just $3.00, as panicked sellers unloaded 15 million shares within a couple of hours.[29]

With algorithms carrying this kind of power, is it any real

wonder we have started to rely on them to tell us what is important and what is not? In early 2009, a small town in France called Eu decided to change its name to one made up of a longer string of text—"Eu-en-Normandie" or "la Ville d'Eu"—because Google searches for "Eu" generated too many results for the European Union, colloquially known as the E.U. Consider also the Twitter-originated concept of incorporating hashtags (#theformula) into messages. A September 2013 comedy sketch on *Late Night with Jimmy Fallon* demonstrated how ill-suited the whole hashtag phenomenon is for meaningful communication in the real world. ("Check it out," says one character, "I brought you some cookies. #Homemade, #OatmealRaisin, #ShowMeTheCookie.") But of course the idea of hashtags isn't to better explain ourselves to other people, but rather to allow us to modify our speech in a way that makes it more easily recognized and distributed by Twitter's search algorithms.

When the content of our speech is not considered relevant, it can prompt the same kind of reaction as a dating website's algorithms determining there is no one well matched with you. During the Occupy Wall Street protests, many participants and supporters used Twitter to coordinate, debate and publicize their efforts. Even though the protests gained considerable media coverage, however, the term failed to "trend" according to Twitter's algorithms—referring to the terms Twitter shows on its home page to indicate the most discussed topics, as indexed from the 250 million tweets sent every day. Although #OccupyWallStreet failed to trend, less pressing comedic memes like #WhatYouFindInLadiesHandbags and #ThingsThirstyPeopleDo seemingly had no difficulty showing up during that same time span.

Although Twitter denied censorship, what is interesting about the user outcry is what it says about the cultural role we imbue algorithms with. "It's a signal moment where the trending topics on Twitter are read as being an indication of the importance of different sorts of social actions," says Paul Dourish, professor of informatics at the Donald Bren School of Information and Computer Sciences at the University of California, Irvine. Dourish likens trending to what appearing on the front page of the *New York Times* or the *Guardian* meant at a time when print newspapers were at the height of their power. To put it another way, Twitter's trending algorithms are a technological reimagining of the old adage about trees falling in the woods with no one around to hear them. If a social movement like Occupy Wall Street doesn't register on so-called social media, has it really taken place at all?

This is another task now attributed to algorithms. In the same way that debates in the 20th century revolved around the idea of journalistic objectivity at the heart of media freedom, so too in the 21st century will algorithms become an increasingly important part of the objectivity conversation. Reappropriating free-speech advocate Alexander Meiklejohn's famous observation, "What is essential is not that everyone shall speak, but that everything worth saying shall be said," those in charge of the most relied-upon algorithms are given the jobs of cultural gatekeepers, tasked with deciding what is worth hearing and how to deal with that material which is not. A decision like algorithmically blocking those comments on a news site that have received a high ratio of negative comments versus positive ones might sound like a neat way of countering spam messages, but it also poses profound queries relating to freedom of speech.[30]

What can be troubling about these processes is that bias can be easily hidden. In many cases, algorithms are tweaked on an ongoing basis, while the interface of a particular service might remain the same. "Often it's the illusion of neutrality, rather than the reality of it," says Harry Surden, associate professor at the University of Colorado Law School, whom I introduced in Chapter 3. To put it another way, many of us assume that algorithms are objective until they aren't.

"It's very difficult to have an open and frank conversation about culture and what is valued versus what falls off the radar screen when most people don't have a clear sense of how decisions are being made," says scholar Ted Striphas. "That's not to say that algorithms are undemocratic, but it does raise questions when it comes to the relationship between democracy and culture." Examples of this cultural unknowability can be seen everywhere. In April 2009, more than 57,000 gay-friendly books disappeared from Amazon's sales ranking lists, based on their erroneous categorization as "adult" titles. While that was a temporary glitch, the so-called #amazonfail incident revealed something that users had previously not been aware of: that the algorithm used to establish the company's sales ranks—previously believed to be an objective measurement tool—was purposely designed to ignore books designated as adult titles. Similar types of algorithmic demotion can be seen all over the place. YouTube demotes sexually suggestive videos so that they do not appear on "Most Viewed" pages or the "Top Favorite" home page generated for new users. Each time Facebook changes its EdgeRank algorithm—designed to ensure that the "most interesting" content makes it to users' News Feeds—there is an existential crisis whereby some content becomes more visible, while others are rendered invisible.[31]

As was the case with Google's "auto-correct" situation, sometimes these decisions are not left up to companies themselves, but to governments. It is the same kind of algorithmic tweaking that means that we will not see child pornography appearing in the results of search engines, that dissident political speech doesn't appear in China, and that websites are made invisible in France if they promote Nazism. We might argue about the various merits or demerits of particular decisions, but simply looking at the results we are presented with—without questioning the algorithms that have brought them to our attention—is, as scholar Tarleton Gillespie has noted, a little like taking in all the viewpoints at a public protest, while failing to notice the number of speakers who have been stopped at the front gate.[32]

Organize the World

Near the start of this conclusion, I asked whether everything could be subject to algorithmization. The natural corollary to this query is, of course, the question of whether everything *should* be subject to algorithmization. To use this book's title, is there anything that should not be made to subserve The Formula? That is the real billion-dollar question—and it is one that is asked too little.

In this book's introduction, I quoted Bill Tancer, writing in his 2009 book *Click: What We Do Online and Why It Matters*, about a formula designed to mathematically pinpoint the most depressing week of the year. As I noted, Tancer's concern had nothing to do with the fact that such a formula could possibly exist (that there was such a thing as the quantifiably most depressing week of the year) and everything to do with the fact that he felt the wrong formula had been chosen.[33]

The idea that there are easy answers to questions like "Who am I?," or "Who should I fall in love with?," or "What is art?" is an appealing one in many ways. In Chapter 2, I introduced you to Garth Sundem, the statistician who created a remarkably pre-scient formula designed to predict the breakup rate of celebrity marriages. When I put the question to him of why such formulas engage the general populace as they do, he gave an answer that smacks more of religiosity than it does of Logical Positivism. "I think people like the idea that there are answers," he says. "I do silly equations, based on questions like whether you should go talk to a girl in a bar. But the thought that there may actually be an answer to things that don't seem answerable is extremely attractive." Does he find it frightening that we might seek to quantify everything, reducing it down to its most granular levels—like the company Hunch I discussed in Chapter 1, which claims to be able to answer any consumer-preference question with 80 to 85 percent accuracy, based only on five data points? "Personally, I think the flip side is scarier," Sundem counters. "I think uncertainty is a lot more terrifying than the potential for mathematical certainty. While I was first coming up with formu-las at college, trying to mathematically determine whether we should go to the library to get some work done, deep down in the recesses of our dorky ids I think that what we were saying is that life is uncertain and we were trying to make it more certain. I'm not as disturbed by numbers providing answers as I am by the potential that there might not be answers."

What is it about the modern world that makes us demand easy answers? Is it that we are naturally pattern-seeking creatures, as the statistician Nate Silver argues in *The Signal and the Noise*? Or is there something about the effects of the march of technology that demands the kind of answers only an algorithm can provide?

"[The algorithm does] seem to be a key metaphor for what matters now in terms of organizing the world," acknowledges McKenzie Wark, a media theorist who has written about digital technologies for the last 20 years. "If one thinks of algorithms as processes which terminate and generate a result, there's a moment when the process ceases and you have your answer. If something won't terminate then it probably means that your computer has gone wrong. There's a sense that we [increasingly] structure reality around processes that will yield results—that we've embedded machine logic in the world."

The idea of the black box is one that comes up a lot when discussing algorithms, and it is one that Bruno Latour seizes upon as a powerful metaphor in his work. The black box is, he notes, a term used by cyberneticians whenever a piece of machinery or else a set of commands is too complex. In its place, the black box stands in as an opaque substitute for a technology in which nothing needs to be known other than inputs and outputs. Once opened, it makes both the creators and the users confront the subjective biases and processes that have resulted in a certain answer. Closed, it becomes the embodiment of objectivity: a machine capable of spitting out binary "yes" or "no" answers without further need of qualification. "Insofar as they consider all the black boxes well sealed, people do not, any more than scientists, live in a world of fiction, representation, symbol, approximation, convention," Latour observes. "They are simply *right*." In this vein, we might also turn to Slavoj Žižek's conception of the toilet bowl: a seemingly straightforward technological mechanism through which excrement disappears from our reality and enters another space we phenomenologically perceive to be a messier, more chaotically primordial reality.[34]

It is possible to see some of this thinking in the work of Richard Berk I profiled in Chapter 3. "It frees me up," Berk said of his future crime prediction algorithm. "I don't care whether it makes sense in any kind of causal way." While Berk's comments are designed to get actionable information to predict future criminality, one could argue that by black-boxing the inner workings of the technology, something similar has taken place with the underlying social dynamics. In other areas—particularly as relate to law—a reliance on algorithms might simply justify existing bias and lack of understanding, in the same way that the "filter bubble" effect described in Chapter 1 can result in some people not being presented with certain pieces of information, which may take the form of opportunities.

"It's not just you and I who don't understand how these algorithms work—the engineers themselves don't understand them entirely," says scholar Ted Striphas. "If you look at the Netflix Prize, one of the things the people responsible for the winning entries said over and over again was that their algorithms worked, even though they couldn't tell you *why* they worked. They might understand how they work from the point of view of mathematical principles, but that math is so complex that it is impossible for a human being to truly follow. That troubles me to some extent. The idea that we don't know the world that we're creating makes it very difficult for us to operate ethically and mindfully within it."

How to Stay Human in the World of The Formula

One of the most disconcerting algorithms I came across during my research was the so-called TruthTeller algorithm developed by the *Washington Post* newspaper in 2012, to coincide with that summer's presidential election season. Capable of scanning through political speeches in real time and informing us of when we are being lied to, the TruthTeller is an uncomfortable reminder of both our belief in algorithmic objectivity and our desire for easy answers. In a breathless article, Geek.com described it as "the most robust, automated way to tell whether a politician is lying or not, even more [accurate] than a polygraph test . . . because politicians are so delusional they end up genuinely believing their lies." The algorithm works by using speech recognition technology developed by Microsoft, which converts audio signals into words, before handing the completed transcript over to a matching algorithm to comb through and compare alleged "facts" to a database of previously recorded, proven facts.[35]

Imagine the potential for manipulation should such a technology ever ascend beyond simple gimmickry to enjoy the ubiquity of, for instance, automated spell-checking algorithms. If such a tool was to be implemented within a future edition of MS Word or Google Docs, it is not inconceivable that users may one day finish typing a document and hit a single button—at which point it is auto-checked for spelling, punctuation, formatting and truthfulness. Already there is widespread use of algorithms in academia for sifting through submitted work and pulling up passages that may or may not be plagiarized. These will only become more widespread as natural language processing becomes more intuitive and able to move beyond simple passage comparison to detailed content and idea analysis.

There is no one-size-fits-all answer to how best to deal with algorithms. In some cases, increased transparency would appear to be the answer. Where algorithms are used to enforce laws, for instance, releasing the source code to the general public would both protect against the dangers of unchecked government policy-making and make it possible to determine how specific decisions have been reached. But this approach also wouldn't work in matters of national security, where revealing the inner workings of specific black boxes would enable certain individuals to "game" the system in such a way as to render the algorithm useless.[36]

A not dissimilar paradox can be seen in what happened to Google's Flu Trends algorithm in 2013. Heralded as a breakthrough development thanks to its ability to track the spread of flu through semantic analysis of user searches, Flu Trends ran into an unexpected problem when it received so much media attention that its algorithm began to malfunction. Previously designed to look for searches like "medicine for cough and fever" and assume that these users were sick, what Google found instead was that people were typing in flu-related searches to look for information about Google's own algorithm. Experiencing spikes in its data, Google predicted near-epidemic levels of flu, which then failed to materialize.

The company wound up admitting that while its algorithms might be on top of the flu problem, they were also "susceptible to heightened media coverage." "The lesson here is rich with irony," wrote *InformationWeek* journalist Thomas Claburn when he reported the story. "To effectively assess data from a public source, the algorithm must remain private, or [else] someone will attempt to introduce bias."[37]

Beyond this there is the question of how we stay human in the world of The Formula. After all, if the original dream of hard AI was to create a computer that could believably behave like a human, then the utilitarian opposite of this is to somehow reduce human activity to the point where it is made as predictable as a computer.

We can find many cases where the algorithmization of these profoundly human activities risks losing what makes them special in the first place. As Jaron Lanier argues in his techno-skeptical book, *You Are Not a Gadget*, "Being a person is not a pat formula, but a quest, a mystery, a leap of faith."[38]

But this is also easier said than done in a world increasingly subject to algorithmic proceduring. So how to survive then? Certainly, there is a minority currently engaged in what we might term "algorithmical data jamming," trying to develop tactics to obscure or evade algorithms' attempts to know and categorize them. But to do this means losing out on some of the most valuable tools of the modern age. Since algorithms increasingly define what is relevant, it also means stepping away from many matters of public discourse.

Instead, I propose that users learn more about the world of The Formula, since knowledge of these algorithmic processes is going to be key to many of the important debates going forward—in everything from human relationships to law. For instance, police must establish reasonable suspicion for a "stop and search" to take place, by pointing to "specific and articulable facts which, taken together with rational inferences from those facts, reasonably warrant that intrusion." In this case, does "the algorithm said so" (provided the algorithm is shown to work effectively) provide enough probable cause to carry out such a stoppage?[39]

Similarly, if human relationships with algorithms are not "real," are they at least "real enough"? In 2013, a group of researchers in Spain successfully coded a piece of software designed to mimic the language and attitude of a 14-year-old girl. Negobot—also referred to as the "Virtual Lolita"—is designed to trap potential pedophiles in online chat rooms. Starting in a neutral mode that talks with strangers about ordinary subjects, Negobot goes into "game mode" when a stranger starts communicating in innuendo or with overt sexual overtones, trying to provoke the person on the other end of the conversation to agree to a meet-up. Should such a technology be adopted by law enforcers, it will suggest that feelings toward an algorithm are at least "real enough" to warrant prosecution. As Al Gore has said, "The ability to code and understand the power of computing is crucial to success in today's hyper-connected world."[40]

Most important of all is asking questions—and not expecting simple answers. One of these questions involves not just what algorithms are doing to us—but what they are designed to do in the first place. This is a pressing question, and one that needs to be asked particularly in cases where a service is ostensibly free to users. Words like "relevant" and "newsworthy" are loaded terms that encourage (but often fail to answer) the seemingly obvious follow-up question: "relevant" and "newsworthy" to whom? In the case of a company like Google, the answer is simple: to the company's shareholders, of course. Facebook's algorithms can similarly be viewed as a formula for maintaining and building your friendship circle—but of course the reality is that Facebook's purpose isn't to make you friends, but rather to monetize your social graph through advertising.[41]

Hopefully, this questioning process is starting to happen. A number of researchers working with recommender systems have told me how user expectations have changed in recent years. Where five or ten years ago, people would be happy with any recommendations, today an increasing number want to know *why* these recommendations have been made for them. With asking why we are expected to take things at "interface value" will come the ability to critique the continued algorithmization of everything. Ultimately there are no catchall answers. Our lives would look a lot different—and, most likely, be far worse—without algorithms. But that doesn't mean we stop asking the important questions.

And particularly when easy answers are seemingly involved.

A Note on Author Interviews

(Conducted 2012–2013)

NB: I was in the privileged position of speaking to a large number of individuals while researching *The Formula*. Not everyone is mentioned by name in the text, but below is a list reflecting the individuals whose contributions are worthy of noting:

Vincent Dean Boyce, Steve Carter, Kristin Celello, John Cheney-Lippold, Noam Chomsky, Danielle Citron, Kevin Conboy, Tom Copeman, Paul Dourish, Robert Epstein, Konrad Feldman, Lee Goldman, Graeme Gordon, Jonathan Gottschall, Guy Halfteck, David Hursh, John Kelly, Alexis Kirke, Alec Levenson, Jiwei Li, Benjamin Liu, Sean Malinowski, Lev Manovich, Nick Meaney, Vivienne Ming, George Mohler, John Parr (UK representative for PCM), Giles Pavey, Richard Posner, Ivan Poupyrev, Stephen Ramsay, Emily Riehl, Anna Ronkainen, Adam Sadilek, Matthew Salganik, Lior Shamir, Larry Smarr, Mark Smolinski, Celestino Soddu, Ted Striphas, Garth Sundem, Harry Surden, Francisco Vico, McKenzie Wark, Neil Clark Warren, Robert Wert.

Notes

An Explanation of the Title, and Other Cyberbole

1   Tancer, Bill. *Click: What We Do Online and Why It Matters* (London: HarperCollins, 2009).

Chapter 1: *The Quantified Selves*

1   technologyreview.com/featuredstory/426968/the-patient-of-the -future/.

2   Smarr, Larry. "Towards Digitally Enabled Genomic Medicine: A 10-Year Detective Story of Quantifying My Body." September 2011. lsmarr.calit2.net/repository/092811_Special_Letter,_Smarr.final.pdf.

Gorbis, Marina. *The Nature of the Future: Dispatches from the Socialstructed World* (New York: Free Press, 2013).

3   Bowden, Mark. "The Measured Man." *The Atlantic*, July 13, 2012. theatlantic.com/magazine/archive/2012/07/the-measured-man/ 309018/.

4   quantifiedself.com.

5   750words.com.

6   Nafus, Dawn, and Jamie Sherman. "This One Does Not Go Up to Eleven: The Quantified Self Movement as an Alternate Big Data

Practice" (Review draft). April 2013. quantifiedself.com/wp-content/uploads/2013/04/NafusShermanQSDraft.pdf.

7   Friedman, Ted. *Electric Dreams: Computers in American Culture* (New York: New York University Press, 2005).

8   media.mit.edu/wearables/.

9   Copeland, Douglas. *Generation X: Tales for an Accelerated Culture* (New York: St. Martin's Press, 1991).

10 James, William. *The Principles of Psychology, Vol. 1* (New York: Henry Holt, 1890).

11 Cockerton, Paul. "Tesco Using Minority Report–Style Face Tracking Technology So Ads on Screens Can Be Tailored." *Irish Mirror*, November 4, 2013. irishmirror.ie/news/tesco-using-minority-report-style-face-2674367.

12 Toffler, Alvin. *The Third Wave* (New York: Morrow, 1980).

    Elias, Norbert. *The Civilizing Process* (New York: Urizen Books, 1978).

13 This wave metaphor was not, in itself, new: the German sociologist Norbert Elias had referred to "a wave of advancing integration over several centuries" in his book *The Civilizing Process*, as had other writers over the previous century.

14 Richtel, Matt. "How Big Data Is Playing Recruiter for Specialized Workers." *New York Times*, April 27, 2013. nytimes.com/2013/04/28/technology/how-big-data-is-playing-recruiter-for-specialized-workers.html?_r=0.

15 Kwoh, Leslie. "Facebook Profiles Found to Predict Job Performance." *Wall Street Journal*, February 21, 2012. online.wsj.com/news/articles/SB10001424052970204909104577235474086304212.

16 Bulmer, Michael. *Francis Galton: Pioneer of Heredity and Biometry* (Baltimore: Johns Hopkins University Press, 2003).

17 Pearson, Karl. *The Life, Letters and Labours of Francis Galton* (Cambridge, UK: Cambridge University Press, between 1914 and 1930).

18 Galton, Francis. *Statistical Inquiries into the Efficacy of Prayer* (Melbourne: H. Thomas, Printer, between 1872 and 1880).

19 Gould, Stephen Jay. *The Mismeasure of Man* (New York: Norton, 1981).

20 Dormehl, Luke. "This Algorithm Can Tell Your Life Story Through Twitter." *Fast Company*, November 4, 2013. fastcolabs.com/3021091/this-algorithm-can-tell-your-life-story-through-twitter.

21 Isaacson, Walter. *Steve Jobs* (New York: Simon & Schuster, 2011).

22 Kosinskia, Michal, David Stillwella, and Thore Graepelb. "Private Traits and Attributes Are Predictable from Digital Records of Human Behavior." *PNAS*, vol. 110, no. 15, April 9, 2013. pnas.org/content/110/15/5802.full.

23 Larsen, Noah. "Bloggers Reveal Personalities with Word, Phrase Choice." *Colorado Arts & Sciences Magazine*, January 2011. artsandsciences.colorado.edu/magazine/2011/01/bloggers-word-choice-bares-their-personalities/.

24 Markham, Annette. "The Algorithmic Self: Layered Accounts of Life and Identity in the 21st Century." *Selected Papers of Internet Research*, 14.0, 2013. spir.aoir.org/index.php/spir/article/download/891/466.

25 Streitfeld, David. "Teacher Knows If You've Done the E-Reading." *New York Times*, April 8, 2013. nytimes.com/2013/04/09/technology/coursesmart-e-textbooks-track-students-progress-for-teachers.html?hp&_r=0.

26 Levy, Steven. *In the Plex: How Google Thinks, Works, and Shapes Our Lives* (New York: Simon & Schuster, 2011).

27 Manjoo, Farhad. "The Happiness Machine." *Slate*, January 21, 2013. slate.com/articles/technology/technology/2013/01/google_people_operations_the_secrets_of_the_world_s_most_scientific_human.html.

28 Taylor, Frederick. *The Principles of Scientific Management* (New York: Norton, 1967, 1947).

29 Hogge, Becky. *Barefoot into Cyberspace: Adventures in Search of Techno-Utopia* (Saffron Walden, UK: Barefoot, 2011).

30 O'Connor, Sarah. "Amazon's Human Robots: They Trek 15 Miles a Day Around a Warehouse, Their Every Move Dictated by Computers Checking Their Work. Is This the Future of the British Workplace?" *Daily Mail*, February 28, 2013. dailymail.co.uk/news/article-2286227/Amazons-human-robots-Is-future-British-workplace.html.

31   O'Connor, Sarah. "Amazon Unpacked." *FT Magazine*, February 8,
     2013. ft.com/cms/s/2/ed6a985c-70bd-11e2-85d0-00144feab49a
     .html#slide0.

32   Brownlee, John. "Think Your Office Is Soulless? Check Out This
     Amazon Fulfillment Center." *Fast Company*, July 1, 2013. fastcodesign
     .com/1672939/think-your-office-is-soulless-check-out-this-amazon
     -fulfillment-center#disqus_thread.

33   Rawlinson, Kevin. "Tesco Accused of Using Electronic Armbands
     to Monitor Its Staff." *Independent*, February 13, 2013. independent
     .co.uk/news/business/news/tesco-accused-of-using-electronic
     -armbands-to-monitor-its-staff-8493952.html.

34   This version of the quote was reprinted in a 1960 *Time* magazine
     article. Other retellings have Churchill say, "We shape our dwell-
     ings, and afterwards our dwellings shape us."

35   Dormehl, Luke. "Why We Need 'Decimated Reality' Aggregators."
     Fast Company, June 21, 2013. Available at: fastcolabs.com/3013382/
     why-we-need-decimated-reality-aggregators.

36   Pariser, Eli. *The Filter Bubble: What the Internet Is Hiding from You*
     (New York: Penguin Press, 2011).

37   Graham, Stephen, and Simon Marvin. *Splintering Urbanism:
     Networked Infrastructures, Technological Mobilities and the Urban
     Condition* (London; New York: Routledge, 2001).

38   Graham, Stephen. "Software-Sorted Geographies." *Progress in
     Human Geography,* vol. 29, no. 5, October 2005. dro.dur.ac.uk/
     194/1/194.pdf.

39   Conway, David. "Dynamic Pricing of Electronic Content" patent.
     Application no. 12973121. Filed December 20, 2010. patft.uspto
     .gov/netacgi/ph-Parser?Sect1=PTO1&Sect2=HITOFF&d=PALL&
     p=1&u=%2Fnetahtml%2FPTO%2Fsrchnum.htm&r=1&f=G&l=50&
     s1=8,260,657.PN.&OS=PN/8,260,657&RS=PN/8,260,657.

40   Wuyts, Ann. "Inferring Mood from Your Smartphone Routine." Blog
     post, July 31, 2013. jini.co/blog/2013/07/infering-mood-from-your
     -smartphone-routine/.

41   Machleit, Karen, and Susan Powell Mantel. "Emotional Response
     and Shopping Satisfaction: Moderating Effects of Shopper
     Attributions." *Journal of Business Research*, vol. 54, no. 2,
     November 2001.

42   Heffernan, Virginia. "Amazon's Prime Suspect." *New York Times*, August 6, 2010. nytimes.com/2010/08/08/magazine/08FOB -medium-t.html?_r=0.

43   Hansen, Mark. "Digitizing the Racialized Body, or The Politics of Universal Address." *SubStance*, vol. 33, no. 2, 2004.

44   Deleuze, Gilles, and Félix Guattari. *A Thousand Plateaus: Capitalism and Schizophrenia* (Minneapolis: University of Minnesota Press, 1987).

45   Deleuze, Gilles. "Postscript on the Societies of Control." *October*, vol. 59, Winter 1992.

46   Poole, Steven. "The Digital Panopticon." *New Statesman*, May 29, 2013. newstatesman.com/sci-tech/sci-tech/2013/05/are-you-ready -era-big-data.

47   Anderson, Chris. *The Long Tail* (London: Random House Business, 2006).

48   Ellul, Jacques. *The Technological Society* (New York: Knopf, 1964).

49   Turner, Fred. *From Counterculture to Cyberculture: Stewart Brand, the Whole Earth Network, and the Rise of Digital Utopianism* (Chicago: University of Chicago Press, 2006).

50   Turkle, Sherry. *Life on Screen: Identity in the Age of the Internet* (New York: Simon & Schuster, 1995).

51   Singer, Natasha. "When Your Data Wanders to Places You've Never Been." *New York Times*, April 27, 2013. nytimes.com/2013/04/28/ technology/personal-data-takes-a-winding-path-into-marketers -hands.html.

52   Sussman, Warren. *Culture as History: The Transformation of American Society in the Twentieth Century* (New York: Pantheon Books, 1984).

Chapter 2: *The Match & the Spark*

1   Gale, David, and Lloyd Shapley. "College Admissions and the Stability of Marriage." *American Mathematical Monthly*, vol. 69, 1962.

2   Sundem, Garth, and John Tierney. "From Tinseltown to Splitsville: Just Do the Math." *New York Times*, September 19, 2006.

3   Stendhal. *On Love* (London: Penguin, 1975).

4   Brazier, David. *Love and Its Disappointment: The Meaning of Life, Therapy and Art* (Winchester, UK: O Books, 2009).

5   Mulvey, Laura. "Visual Pleasure and Narrative Cinema," in *Visual and Other Pleasures* (London: Macmillan, 1989).

6   Nietzsche, Friedrich. *Thus Spoke Zarathustra* (New York: Barnes & Noble Classics, 2007).

7   Woods, Judith. "Internet Dating: We Just Clicked." *Telegraph*, March 23, 2012. telegraph.co.uk/women/sex/online -dating/9160622/Internet-dating-We-just-clicked.html.

8   Moskowitz, Eva. *In Therapy We Trust: America's Obsession with Self-Fulfillment* (Baltimore: Johns Hopkins University Press, 2001).

9   Ibid.

10   Slater, Dan. *Love in the Time of Algorithms: What Technology Does to Meeting and Mating* (New York: Current, 2013).

11   The first personal ad on record was published in the *Athenian Mercury* in 1692, around 50 years after the start of the modern newspaper. A sample advertisement—published in 1777—was written by a young lady seeking "A man of fashion, honour, and sentiment, blended with good nature, and noble spirit, such as one would chuse [sic] for her guardian and protector."

12   E-mail exchange with Jaime Rupert, eHarmony's director of corporate communications. December 10, 2013.

13   Kerckhoff, Alan. "Patterns of Homogamy and the Field of Eligibles." *Social Forces*, vol. 42, no. 3, 1964.

14   Finkel, Eli, Paul Eastwick, Benjamin Karney, Harry Reis and Susan Sprecher. "Online Dating: A Critical Analysis from the Perspective of Psychological Science." *Psychological Science in the Public Interest*, vol. 13, no. 3, 2012. faculty.wcas.northwestern.edu/eli-finkel/docu ments/2012_FinkelEastwickKarneyReisSprecher_PSPI.pdf.

15   okcupid.com/about.

16   genepartner.com/index.php/aboutgenepartner.

17   itunes.apple.com/us/app/find-your-facemate/id703849612?mt=8.

18   Rogers, Abby. "Models, Cheaters and Geeks: How 15 Niche Dating Websites Are Helping All Sorts of People Find Love." *Business Insider*, March 15, 2012. businessinsider.com/15-niche-dating-websites -2012-3?op=1#ixzz2nBFQ44r3.

19   Pinker, Steven. *How the Mind Works* (New York: Norton, 1997).

20   Bauman, Zygmunt. *Liquid Love: On the Frailty of Human Bonds* (Cambridge, UK: Polity Press).

21   Schwartz, Barry. *The Paradox of Choice: Why More Is Less* (New York: Ecco, 2004).

22   McDonald, Christie. *The Proustian Fabric: Associations of Memory* (Lincoln: University of Nebraska Press, 1991).

23   Webb, Amy. "Why Data Is the Secret to Successful Dating." *Guardian*, January 28, 2013. guardian.co.uk/news/datablog/2013/jan/28/why-data-secret-successul-dating.

24   Shteyngart, Gary. *Super Sad True Love Story* (New York: Random House, 2010).

25   Pettman, Dominic. *Look at the Bunny: Totem, Taboo, Technology* (Ropley, UK: Zero, 2013).

26   De Botton, Alain. *On Love* (New York: Atlantic Monthly Press, 1993).

27   Iwatani, Yukari. "Love: Japanese Style." *Wired*, June 11, 1998. wired.com/culture/lifestyle/news/1998/06/12899.

28   Gottlieb, Jenna, and Jill Lawless. "New App Helps Icelanders Avoid Accidental Incest." Associated Press, April 18, 2013. salon.com/2013/04/18/new_app_helps_icelanders_avoid_accidental_incest_ap/.

29   Lanier, Jaron. *Who Owns the Future?* (New York: Simon & Schuster, 2013).

30   Hill, Kashmir. "SceneTap Wants to One Day Tell You the Weights, Heights, Races and Income Levels of the Crowd at Every Bar." *Forbes*, September 25, 2012. forbes.com/sites/kashmirhill/2012/09/25/scenetap-wants-to-one-day-use-weight-height-race-and-income-to-help-you-decide-which-bar-to-go-to/.

31   bedposted.com.

32   Grothaus, Michael. "About 9% of You Would Have Sex with a Robot." *Fast Company*, August 7, 2013. fastcolabs.com/3013217/the-forerunners-of-future-sexbots-now.

33   Jastrow, Robert. *The Enchanted Loom: Mind in the Universe* (New York: Simon and Schuster, 1981).

34   deadsoci.al/index.php.

35   ifidie.net/.

36 Coldwell, Will. "Why Death Is Not the End of Your Social Media Life." *Guardian*, February 18, 2013. theguardian.com/media/short cuts/2013/feb/18/death-social-media-liveson-deadsocial.

37 Saletan, William. "Night of the Living Dad." *Slate*, January 12, 2009. slate.com/articles/health_and_science/human_nature/2009/01/night_of_the_living_dad.html.

38 Merleau-Ponty, Maurice. *Phenomenology of Perception: An Introduction* (London: Routledge, 2002).

39 The first conversation I had with my freshly downloaded "demo" Kari ended on a distinctly gold-digger note when my would-be virtual girlfriend observed that our trial time together had expired and opined, "I am so sad. Register me, so that we can keep talking. I love you."

40 Not coincidentally, the world's first chatterbot—conceived by MIT computer scientist Joseph Weizenbaum—was named ELIZA.

41 Shakespeare, Geoff. "The Six Best Girlfriend Substitutes Technology Has to Offer." *Spike*, August 13, 2010. spike.com/articles/myqbz4/the-six-best-girlfriend-substitutes-technology-has-to-offer.

42 When I asked Sergio Parada about the impediment of being unable to consummate a relationship with Kari—thereby making her a less than ideal lover in some ways—he told me about one user who had bought a sex doll that they had then implanted with a microcomputer, running both Kari and voice recognition software. "He had [the doll] sitting on the couch while he was able to talk to her like a real person," Parada says. Recently Parada was approached by a technology company interested in developing "some kind of gadget that works on suction. It plugs into the USB port and comes with a software development kit." The idea, he says, "was to have Kari interact with this gadget that you slip over your member."

43 Richie, Donald, and Roy Garner. *The Image Factory: Fads and Fashions in Japan* (London: Reaktion, 2003).

44 Levy, David. *Love and Sex with Robots* (New York: HarperPerennial, 2008).

Chapter 3: *Do Algorithms Dream of Electric Laws?*

1 Hays, Constance. "What Wal-Mart Knows About Customers' Habits." *New York Times*, November 14, 2004. nytimes.com/2004/11/14/business/yourmoney/14wal.html.

2   Beck, Charlie, and Colleen McCue. "Predictive Policing: What Can We Learn from Walmart and Amazon about Fighting Crime in a Recession?" *Police Chief,* November 2009. policechiefmagazine.org/magazine/index.cfm?fuseaction=display_arch&article_id=1942&issue_id=112009.

3   Bratton, William, and Peter Knobler. *Turnaround: How America's Top Cop Reversed the Crime Epidemic* (New York: Random House, 1998).

4   Ferguson, Andrew. "Predictive Policing and Reasonable Suspicion." May 2, 2012. law.emory.edu/fileadmin/journals/elj/62/62.2/Ferguson.pdf.

5   thenextweb.com/uk/2011/10/14/us-super-cop-and-new-uk-adviser-talks-data-driven-policing-and-putting-cops-on-the-dots/.

6   pandodaily.com/2012/07/09/predpol-brings-big-data-to-law-enforcement-with-1-3-million-round/.

7   Infante, Francesca. "Met's Minority Report: They Use Computer Algorithms to Predict Where Crime Will Happen." *Daily Mail,* September 29, 2013. dailymail.co.uk/news/article-2437206/Police-tackle-burglars-muggers-using-Minority-Report-style-technology-tackle-future-crime.html.

8   "Predictive Policing Comes to the UK." *Statewatch,* March 5, 2013. statewatch.org/news/2013/mar/01uk-predictive-policing.htm.

9   Comte, Auguste, and Ronald Fletcher. *The Crisis of Industrial Civilization: The Early Essays of Auguste Comte* (London: Heinemann Educational, 1974).

10   Bohm, Robert. *A Primer on Crime and Delinquency Theory* (Belmont, Calif.: Wadsworth, 2001).

11   Hacking, Ian. *The Taming of Chance* (Cambridge, UK; New York: Cambridge University Press, 1990).

12   Rafter, Nicole. *The Origins of Criminology: A Reader* (New York: Routledge, 2009).

13   Mlodinow, Leonard. *The Drunkard's Walk: How Randomness Rules Our Lives* (New York: Pantheon Books, 2008).

14   Belt, Elmer, and Louise Darling. *Elmer Belt Florence Nightingale Collection* (San Francisco: Internet Archive, 2009).

15   Danzigera, Shai, Jonathan Levav and Liora Avnaim-Pesso. "Extraneous Factors in Judicial Decisions." *PNAS*, vol. 108, no. 17, April 26, 2011. pnas.org/content/108/17/6889.full.

16   Markoff, John. "Armies of Expensive Lawyers, Replaced by Cheaper Software." *New York Times*, March 4, 2011. nytimes.com/2011/03/05/science/05legal.html.

17   Lev-Ram, Michal. "Apple v. Samsung: (Patent) Trial of the Century." *Fortune*, January 22, 2013. tech.fortune.cnn.com/2013/01/22/apple-samsung-patent-lawsuit/.

18   Dannen, Chris. "The Amazing Forensic Tech Behind the Next Apple, Samsung Legal Dust-Up (and How to Hack It)." *Fast Company*, December 6, 2012. fastcolabs.com/3003732/amazing-forensic-tech -behind-next-apple-samsung-legal-dust-and-how-hack-it.

19   Christensen, Clayton. *Innovator's Dilemma: When New Technologies Cause Great Firms to Fail* (Boston: Harvard Business School Press, 1997).

20   Dawkins, Richard. *The Selfish Gene* (Oxford, UK; New York: Oxford University Press, 1989).

21   Consider, for instance, the way in which divorce cases are presumed—in their very *"Smith v. Smith"* language—to be adversarial, even when this might not be the case at all.

22   Fisher, Daniel. "Silicon Valley Sees Gold in Internet Legal Services." *Forbes*, October 5, 2011. forbes.com/sites/danielfisher/2011/10/05/silicon-valley-sees-gold-in-internet-legal -services/.

23   Casserly, Meghan. "Can This Y-Combinator Startup's Technology Keep Couples Out of Divorce Court?" *Forbes*, April 10, 2013. forbes .com/sites/meghancasserly/2013/04/10/wevorce-y-combinator -technology-divorce-court/.

24   "After Beta Period, Wevorce Software for Making Every Divorce Amicable Is Now Generally Available Nationwide." May 22, 2013. marketwired.com/press-release/after-beta-period-wevorce-software -making-every-divorce-amicable-is-now-generally-available-1793675 .htm.

25   Turkle, Sherry. *Life on Screen: Identity in the Age of the Internet* (New York: Simon & Schuster, 1995).

26   Lohr, Steve. "Computers That See You and Keep Watch Over You."
      New York Times, January 1, 2011. nytimes.com/2011/01/02/
      science/02see.html?pagewanted=all.

27   Hildebrandt, Mireille. "A Vision of Ambient Law," in *Regulating
      Technologies* (Oxford, UK: Hart, 2008).

28   Massey, Ray. "The Car That Stops You Drink-Driving." *Daily Mail*,
      August 4, 2007. dailymail.co.uk/news/article-473040/The-car-stops
      -drink-driving.html.

29   Winner, Langdon. "Do Artifacts Have Politics?" *Daedalus*, vol. 109,
      no. 1, Winter 1980.

30   Caro, Robert. *The Power Broker: Robert Moses and the Fall of New
      York* (New York: Knopf, 1974).

31   Chivers, Tom. "The Story of Google Maps." *Telegraph*, June 4,
      2013. telegraph.co.uk/technology/google/10090014/The-story-of
      -Google-Maps.html.

32   Lardinois, Frederic. "The Next Frontier for Google Maps Is
      Personalization." *TechCrunch*, February 1, 2013. techcrunch
      .com/2013/02/01/the-next-frontier-for-google-maps-is-personal
      ization/.

33   Seefeld, Bernhard. "Meet the New Google Maps: A Map for Every
      Person and Place." Google Maps, May 15, 2013. google-latlong
      .blogspot.co.uk/2013/05/meet-new-google-maps-map-for-every
      .html.

34   A similar idea was proposed at around the same time by Madeline
      Akrich in her essay "The De-Scription of Technical Objects," which
      appears in the book *Shaping Technology/Building Society: Studies in
      Sociotechnical Change* by Wiebe Bijker and John Law.

35   Latour, Bruno. "On Technical Mediation." *Common Knowledge*,
      vol. 3, no. 2, 1994.

36   To extend this argument to its deterministic extreme, we might turn
      to Karl Marx and his assertion in *The Poverty of Philosophy*: "The
      hand-mill gives you society with the feudal lord; the steam-mill
      society with the industrial capitalist." I take more of a social
      constructionist perspective, seeing technological development as
      the interplay of inventors, entrepreneurs, customers and social
      circumstance.

37 Thaler, Richard, and Cass Sunstein. *Nudge: Improving Decisions about Health, Wealth, and Happiness* (New Haven, Conn.: Yale University Press, 2008).

38 Brownsword, Roger. "What the World Needs Now: Techno-Regulation, Human Rights and Human Dignity," in *Global Governance and the Quest for Justice*. Vol. 4: Human Rights. (Oxford, UK: Hart, 2004).

39 Conly, Sarah. *Against Autonomy* (Cambridge, UK: Cambridge University Press, 2013).

40 Lessig, Lawrence. *Code: Version 2.0* (New York: Basic Books, 2006).

41 Holmes, Oliver Wendell. *The Mind and Faith of Justice Holmes: His Speeches, Essays, Letters, and Judicial Opinions* (New York: Modern Library, 1943).

42 Shay, Lisa, Woodrow Hartzog, John Nelson and Gregory Conti. "Do Robots Dream of Electric Laws: An Experiment in the Law as Algorithm," March 29, 2013. rumint.org/gregconti/publications/201303_AlgoLaw.pdf.

43 Reiser, Stanley. *Medicine and the Reign of Technology* (Cambridge, UK; New York: Cambridge University Press, 1978).

44 Gusfield, Joseph. *The Culture of Public Problems: Drinking-Driving and the Symbolic Order* (Chicago: University of Chicago Press, 1981).

45 "Google's Self-Driving Cars Are Safer Than Human Drivers." *Macworld*, August 8, 2012. macworld.com.au/news/googles-self-driving-cars-are-safer-than-human-drivers-67261/#.Uh2-DLyE5eo.

46 Owen, Glen. "Britain Fights EU's 'Big Brother' Bid to Fit Every Car with Speed Limiter." *Daily Mail*, August 31, 2013. dailymail.co.uk/news/article-2408012/Britain-fights-EUs-Big-Brother-bid-fit-car-speed-limiter.html.

47 Moskvitch, Katia, and Richard Fisher. "Penal Code." *New Scientist*, September 7, 2013.

48 Hook, P. "Police Systems to Automatically Detect Crime." *CCTV Today*, March 19–20, 2001.

49 Graham-Rowe, Duncan. "Warning! Strange Behaviour." *New Scientist*, December 11, 1999. newscientist.com/article/mg16422164.700-warning-strange-behaviour.html.

50   Star, Susan. "The Ethnography of Infrastructure." *American Behavioral Scientist*, vol. 43, no. 3, November 1999. bscw.wineme .fb5.uni-siegen.de/pub/nj_bscw.cgi/d759204/11_2_Star _EthnographyOfInfrastructure.pdf.

51   Irons, Meghan. "Caught in a Dragnet." *Boston Globe*, July 17, 2011. boston.com/yourtown/natick/articles/2011/07/17/man_sues _registry_after_license_mistakenly_revoked/?page=1.

52   calegaladvocates.org/news/article.132896-COMPUTER_ GLITCH_LEAVES_CALIFORNIAS_NEEDIEST_MEDICARE_ RECIPIENTS_WITHOUT_BEN.

53   sfexaminer.com/sanfrancisco/medicare-clients-sue-state-over -computer-flub/Content?oid=2155092.

54   Garvey, Meghan. "Net to Snag Deadbeats Also Snares Innocent." *Los Angeles Times*, April 12, 1998. articles.latimes.com/1998/ apr/12/local/me-38538.

55   United States Government Accountability Office. "Terrorist Watch List Screening: Efforts to Help Reduce Adverse Effects on the Public—Report to Congressional Requesters." 2006.

56   Schneier, Bruce. "Why Data Mining Won't Stop Terror." *Wired*, March 9, 2006. wired.com/politics/security/commentary/ securitymatters/2006/03/70357.

57   Graham, S. (2005) "Software-Sorted Geographies," *Progress in Human Geography*.

58   Sweeney, Latanya. "Google Ads, Black Names and White Names, Racial Discrimination, and Click Advertising." *ACM Queue*, vol. 11, no. 3, March 2013. queue.acm.org/detail.cfm?id=2460278.

59   Ananny, Mike. "The Curious Connection Between Apps for Gay Men and Sex Offenders." *The Atlantic*, April 14, 2011. theatlantic.com/ technology/archive/2011/04/the-curious-connection-between-apps -for-gay-men-and-sex-offenders/237340/.

60   Citron, Danielle. "Technological Due Process." *Washington University Law Review*, vol. 85, 2007.

61   Citron, Danielle. "Technological Due Process" (Luncheon Video/ Audio). January 15, 2008. cyber.law.harvard.edu/interactive/events/ luncheon/2008/01/citron.

62   Posner, Richard. "The Role of the Judge in the Twenty-First Century." *Boston University Law Review*, vol. 86, 2006.

63   Radin, Max. "The Theory of Judicial Decision: Or How Judges Think." 1925.

64   Hutcheson Jr., Joseph C. "The Judgment Intuitive: The Function of the 'Hunch' in Judicial Decision." 1929.

65   Ruger, Theodore, Pauline Kim, Andrew Martin and Kevin Quinn. "The Supreme Court Forecasting Project: Legal and Political Science Approaches to Predicting Supreme Court Decisionmaking." *Columbia Law Review*, vol. 104, no. 4, May 2004.

66   Holmes Jr., Oliver Wendell. "The Path of the Law." 1897.

Chapter 4: *The Machine That Made Art*

1   Goldman, William. *Adventures in the Screen Trade: A Personal View of Hollywood and Screenwriting* (New York: Warner Books, 1983).

2   Medavoy, Mike, and Josh Young. *You're Only as Good as Your Next One: 100 Great Films, 100 Good Films, and 100 for Which I Should Be Shot* (New York: Pocket Books, 2002).

3   Johnston, Rich. "Review: *Avatar*—The Most Expensive American Film Ever . . . and Possibly the Most Anti-American One Too." *Bleeding Cool*, December 11, 2009. bleedingcool.com/2009/12/11/review-avatar-the-most-expensive-american-film-ever-and-the-most-anti-american-one-too/.

4   McBride, Joseph. *Steven Spielberg: A Biography* (New York: Simon & Schuster, 1997).

5   An exchange between Spielberg and his dad, remembered by Arnold Spielberg and recorded in Joseph McBride's acclaimed biography of the director, went like this: "I said, 'Steve, you've gotta study math.' He said, 'I don't like it.' He'd ask me to do his chemistry for him. And he would never even *do* the damn chemistry lab, he would just come home and say, 'Dad, I've gotta prepare this experiment.' I'd say, 'You don't have any data there. How am I supposed to tell you what you've done?' So I'd try to reconstruct the experiment for him, I'd come down with some answers. He'd come back [from school] and say, 'Jesus, Dad, you flunked!' "

6   Keegan, Rebecca. "The Legend of Will Smith." *Time*, November 29, 2007. content.time.com/time/magazine/article/0,9171,1689234,00.html.

7   Eells, Josh. "Jennifer Lawrence: America's Kick-Ass Sweetheart."
    Rolling Stone, April 12, 2012. rollingstone.com/movies/news/
    cover-story-excerpt-jennifer-lawrence-20120328.

8   Whatever it proves, this moment has proven a divisive one among
    critics. While some love it, others view it as the weakest moment
    of the film. In an article for the *New Republic*, critic David
    Thompson referred to it as "appalling." Thompson, David. "Schin-
    dler's Girl in the Red Coat Speaks Out." *New Republic*, March 7,
    2013. newrepublic.com/article/112598/schindlers-girl-red-coat
    -speaks-out.

9   Salganik, Matthew, Peter Dodds and Duncan Watts. "Experimental
    Study of Inequality and Unpredictability in an Artificial Cultural
    Market." *Social Psychology Quarterly*, vol. 71, no. 4, December 2008.
    princeton.edu/~mjs3/salganik_dodds_watts06_full.pdf.

10 Huntzicker, William. *The Popular Press, 1833–1865* (Westport,
    Conn.; London: Greenwood Press, 1999).

11 Rutsky, R. L. *High Technē: Art and Technology from the Machine
    Aesthetic to the Post-human* (Minneapolis: University of Minnesota
    Press, 1999).

12 Manovich, Lev. *The Language of New Media* (Cambridge, Mass.:
    MIT Press, 2002).

13 For anyone interested, the formula he came up with was $S(pi + Pii +
    Piii \ldots P)\, \Upsilon = T$, where S equals the sum of the principles (P), Υ
    equals intuition, and T equals artistic creation.

14 Benjamin, Walter. *The Work of Art in the Age of Mechanical Repro-
    duction* (London: Penguin, 2008).

15 Clark, Liat. "2D Photos Translated into 3D-Printed Translucent
    Artworks." *Wired*, May 23, 2013. wired.co.uk/news/archive/2013
    -05/23/3d-printed-touch-photos.

16 Kim, Seung-Chan, Ali Israr, and Ivan Poupyrev. "Tactile Rendering
    of 3D Features on Touch Surfaces." *Proceedings of the 26th Annual
    ACM Symposium on User Interface Software and Technology—UIST
    2013* (2013): 531–38. disneyresearch.com/wp-content/uploads/
    uist-2013-final.pdf.

17 Poupyrev, Ivan. "Researchers Develop Algorithm for Rendering 3-D
    Tactile Features on Touch Surfaces." October 7, 2013. phys.org/
    news/2013-10-algorithm-d-tactile-features-surfaces.html.

18   The term "info-aesthetics" was first coined in the 1950s by Max
      Bense, a German philosopher with a particular interest in art,
      technology and science.

19   Bishop, Todd. "Bill Gates, Nathan Myhrvold Have Another Wild
      Idea: Automatically Generating Video from Text." *Geekwire*,
      August 13, 2013. geekwire.com/2013/gates-myhrvold-crazy-idea
      -autogenerating-video-text/.

20   Ramsay, Stephen. *Reading Machines: Toward an Algorithmic
      Criticism* (Urbana: University of Illinois Press, 2011).

21   Bowden, B. V. *Faster Than Thought: A Symposium on Digital
      Computing Machines* (New York; London: Pitman, 1953).

22   Shamir, Lior, and Jane Tarakhovsky. "Computer Analysis of Art."
      Journal on Computing and Cultural Heritage, vol. 5, no. 2, July 2012.

23   Shamir, Lior. "Computer Analysis Reveals Similarities between the
      Artistic Styles of Van Gogh and Pollock." *Leonardo*, vol. 45, no. 2,
      April 2012.

24   "Vincent Van Gogh and Jackson Pollock: Changing What Art Is."
      www3.dmagazine.com/events/details/Vincent-Van-Gogh-and
      -Jackson-Pollock-Changing-What-Art-Is.

25   Dormehl, Luke. "Should We Teach Literature Students How to
      Analyze Texts Algorithmically?" *Fast Company*, September 3, 2013.
      fastcolabs.com/3016699/should-we-teach-literature-students-how
      -to-analyze-texts-algorithmically.

26   Leonard, Andrew. "How Netflix Is Turning Viewers into Puppets."
      Salon, February 1, 2013. salon.com/2013/02/01/how_netflix_is
      _turning_viewers_into_puppets/.

27   Bianco, Robert. "*House of Cards* Is All Aces." *USA Today*, February
      1, 2013. usatoday.com/story/life/tv/2013/01/31/bianco-review
      -house-of-cards/1880835/.

28   Blakely, Rhys. "Emmy Awards Brings the Computer Algorithm to
      Hollywood." *Times*, September 20, 2013. thetimes.co.uk/tto/arts/
      tv-radio/article3874137.ece.

29   nytimes.com/2013/07/22/business/media/tv-foresees-its-future
      -netflix-is-there.html?pagewanted=all&_r=0.

30   Levy, Steven. "In Conversation with Jeff Bezos: CEO of the
      Internet." *Wired*, December 12, 2011. wired.co.uk/magazine/
      archive/2012/01/features/ceo-of-the-internet/page/2.

31   Dormehl, Luke. "Can Alternate Endings Save the Hollywood
      Blockbuster?" *Fast Company*, July 30, 2013. fastcolabs.com
      /3015037/open-company/can-alternate-endings-save-the
      -hollywood-blockbuster.

32   This idea is backed up by *Wired* editor Chris Anderson's concept of
      the "98 Percent Rule," described in his 2006 book *The Long Tail*.
      In his discussion of how technology is turning mass markets into
      millions of niches, Anderson argues that niche products are now
      within reach economically thanks to digital distribution, and when
      aggregated can still make up a significant market—as Amazon's
      business model has shown.

33   Warman, Matt. "Xbox One Will Track Viewers' TV Habits and
      Reward Them for Watching Ads." *Telegraph*, May 29, 2013. tele
      graph.co.uk/technology/video-games/Xbox/10087148/Xbox-One
      -will-track-viewers-TV-habits-and-reward-them-for-watching-ads
      .html.

34   Lohr, Steve. "Computers That See You and Keep Watch Over You."
      New York Times, January 1, 2011. nytimes.com/2011/01/02/
      science/02see.html?pagewanted=all.

35   Small, David. "Rethinking the Book," in *Graphic Design &
      Reading* (New York: Allworth Press, 2000).

36   There have been several creative attempts to retain the linearity of
      electronic books using a variety of innovative encryption algorithms.
      In 1992, cyberpunk author William Gibson created an "electronic
      novel" called *Agrippa (A Book of the Dead)*, which was mailed to
      readers on a three-and-a-half-inch floppy disk. Once opened, the
      book's text would be displayed for a single time—gradually
      disappearing as the user scrolled down the computer screen.

37   DeRose, Steven. "Structured Information: Navigation, Access and
      Control." April 1995. sunsite.berkeley.edu/FindingAids/EAD/
      derose.html.

38   The Top Grossing Film of All Time, 1 × 1, 2000. http://salavon
      .com/work/TopGrossingFilmAllTime/.

39   Visit hint.fmw/wind/ to see the wind map in action.

40   bewitched.com/windmap.html.

41   Rushkoff, Douglas. *Present Shock: When Everything Happens Now*
      (New York: Current, 2013).

42   Zittrain, Jonathan. "How Amazon Kindled the Bookburners'
      Flames." *Wired*, July 2013. wired.co.uk/magazine/archive/
      2013/07/ideas-bank/how-amazon-kindled-the-bookburners-flames.

43   Lanier, Jaron. *You Are Not a Gadget: A Manifesto* (New York:
      Alfred A. Knopf, 2010).

44   Hume, David, and John Lenz. *Of the Standard of Taste, and Other
      Essays.* (Indianapolis: Bobbs-Merrill, 1965).

45   Coughlan, Alexandra. "Reviewed: Sensing Memory Festival at the
      University of Plymouth." *New Statesman*, February 21, 2013.
      newstatesman.com/culture/music-and-performance/2013/02/
      aural-pill-popping.

46   Morozov, Evgeny. *To Save Everything, Click Here: Technology,
      Solutionism, and the Urge to Fix Problems That Don't Exist* (London:
      Allen Lane, 2013).

47   Levy, David. *Robots Unlimited: Life in a Virtual Age* (Wellesley,
      UK: A. K. Peters, 2006).

48   Adorno, Theodor, and Max Horkheimer. *Dialectic of Enlightenment*
      (New York: Herder and Herder, 1972).

49   Bell, Philip. "Iamus, Classical Music's Computer Composer, Live
      from Malaga." *Guardian*, July 1, 2012. theguardian.com/
      music/2012/jul/01/iamus-computer-composes-classical-music.

50   Ecker, David. "Of Music and Men." *Columbia Spectator*, January 25,
      2013. columbiaspectator.com/2013/01/25/music-and-men.

Conclusion: *Predicting the Future*

1   Meehl, Paul. *Clinical vs. Statistical Prediction: A Theoretical Analysis
     and a Review of the Evidence* (Minneapolis: University of Minnesota
     Press, 1954).

2   Goode, Erica. "Paul Meehl, 83, An Example for Leaders of
     Psychotherapy, Dies." *New York Times*, February 19, 2003. nytimes
     .com/2003/02/19/obituaries/19MEEH.html.

3   "Automated Computer Algorithms Now Carry Out 70% of
     Trades on U.S. Stock Market." *Colors*, no. 85, December 3, 2012.
     colorsmagazine.com/stories/magazine/85/story/algorithms.

4   Gladwell, Malcolm. *Blink: The Power of Thinking Without Thinking* (New York: Little, Brown, 2005).

5   MacCormick, John. *Nine Algorithms That Changed the Future: The Ingenious Ideas That Drive Today's Computers* (Princeton, N.J.: Princeton University Press, 2012).

6   Levy, Frank, and Richard Murnane. *The New Division of Labor: How Computers Are Creating the Next Job Market* (New York: Russell Sage Foundation; Princeton, N.J.: Princeton University Press, 2004).

7   Stone, Brad. *The Everything Store: Jeff Bezos and the Age of Amazon* (New York: Little, Brown, 2013).

8   Bellos, David. *Is That a Fish in Your Ear? Translation and the Meaning of Everything* (New York: Faber and Faber, 2011).

9   Lanier, Jaron. *Who Owns the Future?* (New York: Simon & Schuster, 2013).

10   jay.law.ou.edu/faculty/Jmaute/Lawyering_21st_Century/ Spring%202012%20files/TheFutureofLaw_DarkClouds.pdf.

11   Brynjolfsson, Erik, and Andrew McAfee. *Race Against the Machine: How the Digital Revolution Is Accelerating Innovation, Driving Productivity, and Irreversibly Transforming Employment and the Economy* (Lexington, Mass.: Digital Frontier Press, 2012).

12   Levitt, Theodore. *Marketing Myopia* (Boston: Harvard Business Press, 2008).

13   Gorz, André. *Farewell to the Working Class: An Essay on Post-Industrial Socialism* (London: Pluto Press, 1982).

       Rifkin, Jeremy. *The End of Work: The Decline of the Global Labor Force and the Dawn of the Post-Market Era* (New York: G. P. Putnam's Sons, 1995).

14   Evans, Christopher. *The Mighty Micro* (Sevenoaks, UK: Coronet, 1980).

15   Keim, Brandon. "Nanosecond Trading Could Make Markets Go Haywire." *Wired*, February 16, 2012. wired.com/wiredscience/ 2012/02/high-speed-trading/.

16   bbc.co.uk/news/technology-18427851.

17   Fallows, Deborah. *Search Engine Users*, January 23, 2005. Pew Research Center and American Life Project, pewinternet.org.

18 Vaidhyanathan, Siva. *The Googlization of Everything (and Why We Should Worry)* (Berkeley: University of California Press, 2011).

19 MacCormick, John. *Nine Algorithms That Changed the Future: The Ingenious Ideas That Drive Today's Computers* (Princeton, N.J.: Princeton University Press, 2012).

20 Anderson, Chris. "The End of Theory: The Data Deluge Makes the Scientific Method Obsolete." *Wired*, June 23, 2008. wired.com/ science/discoveries/magazine/16-07/pb_theory.

21 Doctorow, Cary. "How an Algorithm Came Up with Amazon's 'Keep Calm and Rape a Lot' T-Shirt." *BoingBoing*, March 2, 2013. boingboing.net/2013/03/02/how-an-algorithm-came-up-with .html.

22 "Google Sued over Bettina Wulff Search Results." BBC, September 10, 2012. bbc.co.uk/news/technology-19542938.

23 "Life Through Google's Eyes: Do You Fit Your Search Engine Age Profile?" *Huffington Post UK*, January 5, 2013. huffingtonpost .co.uk/2013/05/01/life-through-googles-eyes-video-avatar_n _3190281.html.

24 "The Google Autocomplete Guide to Politicians." *Huffington Post UK*, April 3, 2013. huffingtonpost.co.uk/2013/04/03/google -autocomplete-politicians_n_3007248.html.

25 Morozov, Evgeny. *The Net Delusion: The Dark Side of Internet Freedom* (New York: PublicAffairs, 2011).

Tung, Liam. "Google Ordered to Muzzle Defamatory Autocompletes by German Court." *ZDNet*, May 15, 2013. zdnet.com/google -ordered-to-muzzle-defamatory-autocompletes-by-german-court -7000015406/.

26 Tuchman, Gaye. "Objectivity as Strategic Ritual: An Examination of Newsmen's Notions of Objectivity." *American Journal of Sociology*, vol. 77, no. 4, January 1972. https://umdrive.memphis.edu/ cbrown14/public/Mass%20Comm%20Theory/Week%2012%20 Encoding/Tuchman%201972.pdf.

27 Mayer, Marissa. "Google I/O '08 Keynote Address." June 5, 2008.

28 Slavin, Kevin. "How Algorithms Shape Our World." *TED Talk*, 2011. youtube.com/watch?v=ENWVRcMGDoU.

Thomas, W. I., and D. S. Thomas. *The Child in America: Behavior Problems and Programs* (New York: Knopf, 1928).

29 "United Airlines Stock Decline & the Power of Google." *OneUp-Web.* oneupweb.com/blog/united_airlines/.

30 Meiklejohn, Alexander. *Political Freedom: The Constitutional Powers of the People* (New York: Harper, 1960).

31 Resende, Patricia. "YouTube Clamps Down on Sexual Content." *NewsFactor*, December 3, 2008. newsfactor.com/news/You Tube-Gets-Tough-on-Sleaze/story.xhtml?story_id=0030009XC5M6.

32 Gillespie, Tarleton. "The Relevance of Algorithms," in *Media Technologies: Paths Forward in Social Research* (Cambridge, Mass.: MIT Press, 2013).

33 Tancer, Bill. *Click: What We Do Online and Why It Matters* (London: HarperCollins, 2009).

34 Latour, Bruno. *Science in Action: How to Follow Scientists and Engineers Through Society* (Cambridge, Mass.: Harvard University Press, 1987).

35 truthteller.washingtonpost.com/about/.

Raby, Mark. "Truth Teller Algorithm Can Fact-Check Politicians in Real Time." *Geek*, January 30, 2013. geek.com/articles/news/truth-teller-can-fact-check-politicians-in-real-time-20130130/.

36 Citron, Danielle. "Technological Due Process." *Washington University Law Review*, vol. 85, 2007.

37 Claburn, Thomas. "How Google Flu Trends Blew It." *InformationWeek*, October 25, 2013. informationweek.com/applications/how-google-flu-trends-blew-it/d/d-id/1112081?

38 Lanier, Jaron. *You Are Not a Gadget: A Manifesto* (New York: Alfred A. Knopf, 2010).

Badiou, Alain. *In Praise of Love* (London: Serpent's Tail, 2012).

39 Ferguson, Andrew. "Predictive Policing and Reasonable Suspicion," May 2, 2012. law.emory.edu/fileadmin/journals/elj/62/62.2/Ferguson.pdf.

40 Lovejoy, Ben. "Apple Offers Free 1-Hour Computer Science Workshops for Kids & Teens." *9to5Mac*, December 9, 2013. 9to5mac.com/2013/12/09/apple-oofers-free-1-hour-computer-science-workshops-for-kids-teens/.

Grothaus, Michael. "Does the 'Lolita Bot' Help Catch Online
Predators or Create More of Them?" *Fast Company*, July 16, 2013.
fastcolabs.com/3013217/the-forerunners-of-future-sexbots-now.

41   Žižek, Slavoj. *Living in the End Times* (London; New York: Verso,
2010).

Index